Editorial Policy

for the publication of monographs

In what follows all references to monographs are applicable also to multiauthorship volumes such as seminar notes.

§ 1. Lecture Notes aim to report new developments - quickly, informally, and at a high level. Monograph manuscripts should be reasonably self-contained and rounded off. Thus they may, and often will, present not only results of the author but also related work by other people. Furthermore, the manuscripts should provide sufficient motivation, examples, and applications. This clearly distinguishes Lecture Notes manuscripts from journal articles which normally are very concise. Articles intended for a journal but too long to be accepted by most journals usually do not have this "lecture notes" character. For similar reasons it is unusual for Ph.D. theses to be accepted for the Lecture Notes series.

§ 2. Manuscripts or plans for Lecture Notes volumes should be submitted (preferably in duplicate) either to one of the series editors or to Springer-Verlag, New York. These proposals are then refereed. A final decision concerning publication can only be made on the basis of the complete manuscript, but a preliminary decision can often be based on partial information: a fairly detailed outline describing the planned contents of each chapter, and an indication of the estimated length, a bibliography, and one or two sample chapters - or a first draft of the manuscript. The editors will try to make the preliminary decision as definite as they can on the basis of the available information.

§ 3. Final manuscripts should be in English. They should contain at least 100 pages of scientific text and should include

- a table of contents;
- an informative introduction, perhaps with some historical remarks: it should be accessible to a reader not particularly familiar with the topic treated;
- a subject index: as a rule this is genuinely helpful for the reader.

Lecture Notes in Statistics 82

Edited by J. Berger, S. Fienberg, J. Gani, K. Krickeberg, I. Olkin, and B. Singer

A. P. Korostelev
A. B. Tsybakov

Minimax Theory of Image Reconstruction

Springer-Verlag
New York Berlin Heidelberg London Paris
Tokyo Hong Kong Barcelona Budapest

A. P. Korostelev
Institute for Systems Studies
Prospeck 60-letija Octjabarja 9
117312 Moscow
Russia

A. B. Tsybakov
Institute for Problems of Information Transmission
Ermolovoy Street 19
101447 Moscow GSP - 4
Russia

Mathematics Subject Classification: 68U10, 62G05

Library of Congress Cataloging-in-Publication Data
Korostelev, A. P. (Aleksandr Petrovich)
Minimax theory of image reconstruction / A. Korostelev, A. Tsybakov)
p. cm. -- (Lecture notes in statistics ; 82)
Includes bibliographical references and indexes.
ISBN-13: 978-0-387-94028-1
1. Image processing--Digital techniques. 2. Image processing--Statistical methods. 3. Image reconstruction. 4. Chebyshev approximation. I. Tsybakov, A. (A. B.) II. Title. III. Series: Lecture notes in statistics (Springer-Verlag) ; v. 82.
TA1637.K67 1993
621.367--dc20 93-18028

Printed on acid-free paper.

Camera ready copy provided by the authors.

9 8 7 6 5 4 3 2 1

ISBN-13: 978-0-387-94028-1 e-ISBN-13: 978-1-4612-2712-0
DOI: 10.1007/978-1-4612-2712-0

PREFACE

There exists a large variety of image reconstruction methods proposed by different authors (see e.g. Pratt (1978), Rosenfeld and Kak (1982), Marr (1982)). Selection of an appropriate method for a specific problem in image analysis has been always considered as an art. How to find the image reconstruction method which is optimal in some sense? In this book we give an answer to this question using the asymptotic minimax approach in the spirit of Ibragimov and Khasminskii (1980a,b, 1981, 1982), Bretagnolle and Huber (1979), Stone (1980, 1982). We assume that the image belongs to a certain functional class and we find the image estimators that achieve the best order of accuracy for the worst images in the class. This concept of optimality is rather rough since only the order of accuracy is optimized. However, it is useful for comparing various image reconstruction methods. For example, we show that some popular methods such as simple linewise processing and linear estimation are not optimal for images with sharp edges. Note that discontinuity of images is an important specific feature appearing in most practical situations where one has to distinguish between the "image domain" and the "background".

The approach of this book is based on generalization of nonparametric regression and nonparametric change-point techniques. We discuss these two basic problems in Chapter 1. Chapter 2 is devoted to minimax lower bounds for arbitrary estimators in general statistical models. These are the main tools applied in the book. In Chapters 1 and 2 the exposition is mostly tutorial. They present a general introduction to nonparametric estimation theory. To prove the theorems some nonstandard methods are chosen which are, in our opinion, simple and transparent. Chapters 3-9 contain mostly the new results, and the reader who is familiar with nonparametric estimation background may proceed to them directly.

The working example that we study in detail is the two-dimensional binary image of "boundary fragment" type. Roughly speaking, it is a small piece of discontinuous image containing the discontinuity curve (the boundary). Imposing some smoothness restrictions on the boundary we find the minimax rates and optimal estimators for boundary fragments. This is the main message of Chapters 3 and 4. Various extensions are discussed in Chapter 5. Some proofs in Chapter 5 and in the following chapters are not detailed. Simple but technical steps of the proofs are sometimes left to the reader.

Chapter 6 deals with the simplified image reconstruction procedures, namely with linewise and linear processing. It is shown that linewise processing can be organized in such a way that it has the optimal rate of convergence in minimax sense. Linear procedures, however, are proved to be suboptimal.

In Chapters 7-9 we discuss some further issues related to the basic image reconstruction problem, namely: the estimation of support of a density, the estimation of image functionals, image estimation from indirect observations, the stochastic tomography setup. For all these problems we derive the minimax rates of convergence and construct the optimal estimators.

One of the points raised in the book is the choice of design in image reconstruction. This point is often ignored since in practice the simplest regular grid design has no competitors. We show that the choice of design is important in image analysis: some randomized designs allow to improve substantially the accuracy of estimation as compared to the regular grid design.

We also consider in brief some parametric imaging problems (Section 1.9, Section 8.2). For parametric case we refer to the continuous-"time" models where the image is supposed to be a solution of a stochastic differential equation. This makes the proofs more concise. Readers who are not familiar with stochastic differential equations may easily skip this part of the book.

Our attitude in this book is to prove the results under the simplest assumptions which still allow to keep the main features of a particular problem. For example, we often assume that the random errors are independent identically distributed Gaussian. Generalizations are mainly given without proofs.

Some words about the notation. We use the small letters c, c_i, $i=1,2,\ldots$, and letter λ (possibly, with indices) to denote positive constants appearing in the proofs. This notation is kept only inside a chapter, so that in different chapters c_i may be different. The constants C_0, C_1 are reserved for the lower and upper bounds of minimax risks respectively. They are different in different theorems.

The work on this book was strongly influenced by the ideas of I.A.Ibragimov and R.Z.Khasminskii and stimulated by the discussions at the seminar of M.S.Pinsker and R.Z.Khasminskii in the Institute for Problems of Information Transmission in Moscow. We would like to thank E.Mammen and A.S.Nemirovskii for helpful personal discussions and suggestions. We are grateful to W.Härdle B.Park, M.Rudemo and B.Turlach who made important remarks that helped much to improve the text.

A.P.Korostelev

A.B.Tsybakov

CONTENTS

CHAPTER 1. NONPARAMETRIC REGRESSION AND CHANGE-POINT PROBLEMS

1.1. INTRODUCTION

In this chapter we describe the basic mathematical tools used for two statistical problems: that of nonparametric regression and of change-point estimation. The techniques of this chapter apply in several ways for the construction and analysis of image estimators. These applications will first appear in Chapter 4. The purpose of this chapter is to give a simple introduction to nonparametric regression and to change-point estimation in a self-sufficient form. We do not propose an overview of all estimation techniques available for these problems. For nonparametric regression we study only an important class of locally-polynomial estimators which contains the popular kernel estimator as a special case, and the class of piecewise-polynomial estimators. For the change-point problem we consider the maximum likelihood estimator. The results that we prove in this chapter are related mainly to the convergence rates of the estimators.

1.2. THE NONPARAMETRIC REGRESSION PROBLEM

Let X, Y be random variables and let $(X_1, Y_1), \ldots, (X_n, Y_n)$ be n independent pairs of random variables having the same distribution as (X, Y). The problem of <u>nonparametric regression with random design</u> consists in estimation of the function

$$f(x) = E(Y|X=x)$$

based on observations $(X_1, Y_1), \ldots, (X_n, Y_n)$. Here $E(Y|X=x)$ is the conditional expectation of Y for fixed $X=x$. Random variables X_i are called <u>design points</u>. The <u>design</u> is defined as a collection $\mathfrak{X}=(X_1, \ldots, X_n)$. Note that

$$(1.1) \qquad Y_i = f(X_i) + \xi_i, \quad i = 1, \ldots, n,$$

where ξ_i are independent random variables such that $E(\xi_i|X_i) = 0$.

Moreover, we assume in this book that $(\xi_1,\dots,\xi_n)$ is independent of $\mathcal{X}$.

The word "nonparametric" indicates that nothing is known a priori about a parametric form of the function f. In other words, the statistician does not know whether f(x) is a member of any parametric family of the form $\{g(x,\theta),\ \theta\in\Theta\}$ where $g(\cdot,\cdot)$ is a given function and Θ is a given subset of a finite-dimensional space. For example, a priori information about f can be either of the following

(i) f is a measurable function

(ii) f is a continuous function

(iii) f is a convex function

(iv) f is a smooth function with known number of continuous derivatives.

Of course, not much can be expected from regression estimates if (i) or (ii) hold (only some kind of consistency, see e.g. Stone (1977)). Condition (iii) shows a special type of a priori information that we don't consider here. In the following we concentrate on the case (iv) which is rather general and at the same time specified enough to guarantee certain rates of convergence for regression estimators. Formally we write (iv) as: $f\in\Sigma(\beta,L)$ where β, L are positive constants, and $\Sigma(\beta,L)$ is the class of functions $g(x)$, $x\in[0,1]$, such that the kth derivative of g exists and satisfies the Hölder condition:

$$\Sigma(\beta,L) = \left\{ g(x):\ |g^{(k)}(x)-g^{(k)}(x')| \le L|x-x'|^{\beta-k},\ x,x'\in[0,1] \right\}$$

($k=\lfloor\beta\rfloor$ is the maximal integer such that $k<\beta$). If $\beta\ge1$ is an integer, then $\Sigma(\beta,L)$ contains continuous functions having the Lipschitzian $(\beta-1)$th derivative.

Sometimes it is necessary to assume that the design points X_i are nonidentically distributed or nonrandom. The simplest example that we shall often refer to is the following. Assume that $X_i=i/n$, $i=1,\dots,n$, and ξ_i are i.i.d. random variables with $E(\xi_i) = 0$. The problem consists in estimation of the function f from observations $Y_1,\dots,Y_n$ satisfying (1.1). It is called <u>nonparametric regression problem with regular grid design</u> (the <u>regular grid</u> on [0,1] with step 1/n is, by definition, the set

$\{\frac{1}{n},\frac{2}{n},\ldots,1\})$. The design is called deterministic if X_i's are fixed (nonrandom). Regular grid design is a special example of deterministic design.

1.3. KERNEL ESTIMATORS

The regression estimators studied in this chapter are linear, i.e. they have the following general form

$$\hat{f}_n(x) = \sum_{i=1}^{n} Y_i W_{ni}(x)$$

where the weights $W_{ni}(x) = W_{ni}(x,X_1,\ldots,X_n)$ only may differ.

Kernel regression estimators were first proposed independently by Nadaraya and Watson in 1964 who considered the case of random design. The Nadaraya-Watson kernel estimator is defined as

$$\hat{f}_n(x) = \sum_{i=1}^{n} Y_i K((X_i-x)/h_n) / \sum_{i=1}^{n} K((X_i-x)/h_n) \tag{1.2}$$

where $\{h_n\}$ is a sequence of positive numbers, $h_n\to 0$, and K is a function on R^1 satisfying

$$\lim_{|u|\to\infty} |u||K(u)| = 0, \tag{1.3}$$

$$\int_{-\infty}^{\infty} |K(u)|du < \infty, \qquad \sup_{u\in R^1} |K(u)| < \infty.$$

In (1.2) and later we set by definition $a/0 = 0$ for $a\in R^1$.

The positive number h_n is called bandwidth and the function K satisfying (1.3) is called kernel.

Definition (1.2) shows that the Nadaraya-Watson estimator is linear one with the explicit expression for the weights:

$$W_{ni}(x) = K((X_i - x)/h_n) / \sum_{i=1}^{n} K((X_i - x)/h_n)$$

Usually K is normalized, so that

$$(1.4) \qquad \int_{-\infty}^{\infty} K(u)\, du = 1.$$

Condition (1.4) is not a restriction since the estimator (1.2) is invariant under multiplication of a kernel by a nonzero factor.

Examples of kernels are $K(u)=(1/2)I\{|u|\le 1\}$ (rectangular kernel), $K(u)=(3/4)(1-u^2)I\{|u|\le 1\}$ (Epanechnikov kernel), $K(u)=(1/\sqrt{2\pi})\exp(-u^2/2)$ (Gaussian kernel). (Here and later $I\{\cdot\}$ denotes the indicator function.) Usually K is chosen to be an even function.

The idea of kernel estimation is simple. Let us explain it using the example of rectangular kernel. In this case the estimator (1.2) is the running mean: the estimator at point x is the average of observations Y_i such that X_i are in the "window" $[x-h_n, x+h_n]$. If $h_n \to \infty$ then the estimator tends to $n^{-1}\sum_{i=1}^{n} Y_i$, the average of all observations, and thus for nonconstant functions f the bias becomes large. If h_n is very small (less than the pairwise distances between design points X_i) then the estimator reproduces the data: $\hat{f}_n(X_i)=Y_i$. In this extremal case the variance becomes high (especially when Y_i are spiky). Note that for the kernel with unbounded support, e.g. for the Gaussian kernel, there is no explicit "window", and all observations are averaged, although with different weights. The weights are decreasing as the distance between X_i and x increases.

Kernel estimators show nice asymptotic behavior under the appropriate choice of h_n. As we mentioned already, the increase of h_n tends to increase the bias of estimator (1.2), while the small values of h_n lead to higher variance. The balance between bias and variance results in an optimal value of h_n. This will be discussed in more detail later.

The Nadaraya-Watson estimator is closely related to the Parzen-Rosenblatt kernel density estimator. Let $X_1,\ldots,X_n$ be i.i.d. random variables with common density $\mu(x)$,$x\in R^1$. The Parzen-Rosenblatt estimator of $\mu(x)$ is defined as

$$\hat{\mu}_n(x) = (nh_n)^{-1} \sum_{i=1}^{n} K((X_i-x)/h_n).$$

It is possible to derive the Nadaraya-Watson estimator (1.2) from the Parzen-Rosenblatt estimator. For this one needs the two-dimensional version of $\hat{\mu}_n$. If $X_i,x\in R^2$, and $X_i=(X_{i1},X_{i2})$, $x=(x_1,x_2)$ where X_{ij},x_j ,$j=1,2$, are components of the vectors X_i,x then the kernel estimate of two-dimensional density of X_i's is defined by

$$(1.5) \qquad \hat{\mu}_n(x_1,x_2) = (nh_n^2)^{-1} \sum_{i=1}^{n} K_1((X_{i1}-x_1)/h_n)K_2((X_{i2}-x_2)/h_n)$$

where K_1,K_2 are some kernels.

Now assume that there exists a joint density $\mu(x,y)$ of random variables (X,Y). Then the conditional expectation is

$$(1.6) \qquad f(x) = \int y\ \mu(x,y)\ dy \Big/ \int \mu(x,y)\ dy$$

To get an estimator of $f(x)$ let us replace $\mu(x,y)$ in (1.6) by its estimator $\hat{\mu}_n(x,y)$:

$$(1.7) \qquad \hat{f}_n(x) = \int y\ \hat{\mu}_n(x,y)\ dy \Big/ \int \hat{\mu}_n(x,y)\ dy.$$

1.3.1. PROPOSITION. Let $\hat{\mu}_n(x,y)$ be the kernel estimator (1.5) where $K_1=K$ is a kernel as in (1.3), and K_2 satisfies the additional condition

$$(1.8) \qquad \int_{-\infty}^{\infty} K_2(u)\, du = 1, \qquad \int_{-\infty}^{\infty} uK_2(u)\, du = 0.$$

Under this choice of $\hat{\mu}_n$ the estimator (1.7) is the Nadaraya-Watson estimator.

Proof.

$$\int y\, \hat{\mu}_n(x,y)\, dy = (nh_n)^{-1} \sum_{i=1}^{n} \left(\int \big((y-Y_i)/h_n\big) K_2((Y_i-y)/h_n)dy \; + \right.$$

$$\left. + \; Y_i \int (1/h_n) K_2((Y_i-y)/h_n)dy \right) K((X_i-x)/h_n)$$

where in view of (1.8) the first integral in the right-hand side vanishes, and the second integral equals to 1. Hence

$$\int y\, \hat{\mu}_n(x,y)\, dy = (nh_n)^{-1} \sum_{i=1}^{n} Y_i K((X_i-x)/h_n).$$

Similarly

$$\int \hat{\mu}_n(x,y)\, dy = (nh_n)^{-1} \sum_{i=1}^{n} K((X_i-x)/h_n). \qquad \square$$

1.4. LOCALLY-POLYNOMIAL ESTIMATORS

Here we introduce a larger class of estimators that contains the Nadaraya-Watson estimator as a special case. To start with, note that the Nadaraya-Watson estimator can be expressed as the solution of the following minimization problem

$$\min_{\theta\in R^1} \sum_{i=1}^{n} (Y_i-\theta)^2 K((X_i-x)/h_n).$$

Therefore this estimator can be viewed as a weighted least squares fit of a constant to an unknown function f. The weights are determined by the kernel and they are small or vanishing outside a neighborhood of x (in other words, the constant fit is local). The bandwidth h_n controls the size of a neighborhood. To generalize this approach one may consider local approximations of f(x) by some non-constant functions. The most important example is the local least squares approximation by polynomials (Stone (1977,1980,1982), Katkovnik (1979,1983,1985), Cleveland (1979), Härdle (1990)). Assume that f can be expanded in Taylor series, and

$$f(z) \approx f(x) + f'(x)(z-x) + \dots + f^{(k)}(x)(z-x)^k/k! =$$

$$=\langle\theta(x),\mathbf{F}((z-x)/h_n)\rangle$$

for z close to x and for some integer $k\geq 0$. Here

$$\theta(x) = (f(x),h_n f'(x),\dots,h_n^k f^{(k)}(x))^T,$$

$$\mathbf{F}(z) = (1,z,z^2/2,\dots,z^k/k!)^T,$$

and $\langle\theta,\mathbf{F}\rangle$ denotes the inner product, i.e. for $\theta = (\theta_0,\dots,\theta_k)^T$, $\mathbf{F} = (\mathbf{F}_0,\dots,\mathbf{F}_k)^T$ we have

$$\langle\theta,\mathbf{F}\rangle = \sum_{j=0}^{k} \theta_j \mathbf{F}_j.$$

Define the estimator for $\theta(x)$ as follows:

$$(1.9)\qquad \hat{\theta}_n(x) = \operatorname*{argmin}_{\theta\in R^{k+1}} \sum_{i=1}^{n} (Y_i-\langle\theta,\mathbf{F}((X_i-x)/h_n)\rangle)^2 K((X_i-x)/h_n).$$

It is convenient to write the vector $\hat{\theta}_n(x)$ in componentwise form as

$$\hat{\boldsymbol\theta}_n(x) = (\hat{\theta}_{n,0}(x), \hat{\theta}_{n,1}(x)h_n, \ldots, \hat{\theta}_{n,k}(x)h_n^k)^T$$

1.4.1.DEFINITION. The <u>locally-polynomial</u> <u>estimator</u> <u>of</u> <u>order</u> k (or LPE(k)) for a regression function at a fixed point x is the first component of the vector $\hat{\boldsymbol\theta}_n(x)$: $\hat{f}_n(x) = \hat{\theta}_{n,0}(x)$.

Note that $\mathbf{F}(0) = (1,0,\ldots,0)^T$. Hence the locally-polynomial estimator can be written as

$$\hat{f}_n(x) = \langle\hat{\boldsymbol\theta}_n(x), \mathbf{F}(0)\rangle. \tag{1.10}$$

For the rest of this section we suppose that K is a nonnegative function.

Introducing the notation

$$\bar{Y}_i = Y_i K^{1/2}((X_i-x)/h_n),$$

$$\bar{\mathbf{F}}_i = \mathbf{F}((X_i-x)/h_n)K^{1/2}((X_i-x)/h_n)$$

we can write (1.9) as

$$\hat{\boldsymbol\theta}_n(x) = \underset{\boldsymbol\theta\in R^{k+1}}{\operatorname{argmin}} \sum_{i=1}^{n} (\bar{Y}_i - \langle\boldsymbol\theta, \bar{\mathbf{F}}_i\rangle)^2.$$

This shows that $\hat{\boldsymbol\theta}_n(x)$ is a standard least squares estimator for each x, and thus it satisfies the system of normal equations

$$\mathcal{B}\,\hat{\boldsymbol\theta}_n(x) = \mathbf{a} \tag{1.11}$$

where $\mathcal{B}$ is the following matrix:

$$\mathcal{B} = (nh_n)^{-1} \sum_{i=1}^{n} \mathbf{F}((X_i-x)/h_n)\mathbf{F}^T((X_i-x)/h_n)K((X_i-x)/h_n),$$

and $\mathbf{a}$ is the following vector:

$$\mathbf{a} = (nh_n)^{-1} \sum_{i=1}^{n} Y_i \mathbf{F}((X_i-x)/h_n)K((X_i-x)/h_n).$$

It will be shown in Section 1.6 that under standard assumptions on K and on X_i's the matrix $\mathcal{B}$ is non-singular (positive definite) for n large enough. In the case of nonsingular $\mathcal{B}$

$$(1.12) \qquad \hat{\theta}_n(x) = \mathcal{B}^{-1}\mathbf{a}.$$

Thus, the locally-polynomial estimator is a linear one and it can be represented as follows

$$\hat{f}_n(x) = \sum_{i=1}^{n} Y_i W_{ni}^{*}(x)$$

where

$$W_{ni}^{*}(x) = \langle \mathbf{F}(0), \mathcal{B}^{-1}(nh_n)^{-1}\mathbf{F}((X_i-x)/h_n)K((X_i-x)/h_n)\rangle$$

1.4.2. PROPOSITION.

(i) The Nadaraya-Watson estimator with nonnegative kernel K is LPE(0).

(ii) Assume that the errors $\xi_i \equiv 0$ (resp. $Y_i = f(X_i)$), f is a polynomial of order less or equal to $\boldsymbol{k}$, $\hat{f}_n$ is LPE($\boldsymbol{k}$) and the matrix $\mathcal{B}$ is nonsingular. Then for all $x \in [0,1]$

$$\hat{f}_n(x) = f(x) \qquad \text{(a.s.)},$$

or, equivalently,

$$\sum_{i=1}^{n} f(X_i) W_{ni}^{*}(x) = f(x) \qquad \text{(a.s.)}$$

(iii) Assume that f is a polynomial of order less or equal to $\boldsymbol{k}$, $\hat{f}_n$ is LPE($\boldsymbol{k}$) and the matrix $\mathcal{B}$ is nonsingular. If the design is deterministic then $\hat{f}_n$ is unbiased:

$$E_f(\hat{f}_n(x)) = f(x), \quad \forall\ x\in[0,1].$$

Here and later in this chapter E_f denotes expectation with respect to the joint distribution of observations (X_i,Y_i), $i=1,\ldots,n$, satisfying (1.1).

Proof. (i) The locally-polynomial estimator of order 0 has an explicit expression which coincides with (1.2).

(ii) If $Y_i=f(X_i)$ is a polynomial of order $\le k$ then for any x

$$f(X_i) = f(x) + f'(x)(X_i-x) + \ldots + f^{(k)}(x)(X_i-x)^k/k! =$$

$$=\langle\theta(x),\mathbf{F}((X_i-x)/h_n)\rangle,$$

and (1.9) has the form

$$\hat{\theta}_n(x)=\underset{\theta\in R^{k+1}}{\operatorname{argmin}}\sum_{i=1}^{n}(\langle\theta(x),\mathbf{F}((X_i-x)/h_n)\rangle-\langle\theta,\mathbf{F}((X_i-x)/h_n)\rangle)^2K((X_i-x)/h_n).$$

This problem has a unique solution $\hat{\theta}_n(x)=\theta(x)$ since the matrix $\mathcal{B}$ is non-singular.

(iii) Follows from (ii): taking expectations of both sides of (1.12) we find that $E_f(\hat{f}_n(x))$ has the same form as $\hat{f}_n(x)$ in (ii).

The property 1.4.2(i) can be in some sense inverted, at least for the case of deterministic design (Müller(1987)): there exists a Nadaraya-Watson estimator (with a kernel depending on k) which is asymptotically equivalent to locally-polynomial estimator.

1.5. PIECEWISE-POLYNOMIAL ESTIMATORS

Another nonparametric regression estimator that will appear several times in this book is a piecewise-polynomial one. It is based on the same idea as the locally-polynomial estimator, although, the polynomial fits are taken in bins of fixed length δ_n rather than in h_n-neighborhood of a current point x.

The simplest example of piecewise-polynomial estimator is piecewise-constant estimator, or regressogram. The value of the regressogram in each bin equals to the average of observations Y_i such that X_i are in the bin. The formal definition of regressogram is the following. Let $\delta_n \to 0$ be a positive sequence (to simplify the notation assume without loss of generality that $M = M_n = 1/\delta_n$ is an integer). Denote $u_\ell = \ell\delta_n$, $\ell=0,\ldots,M$, and divide the interval $[0,1]$ into M subintervals (bins) of the form $U_1=[0,u_1)$, $U_2=[u_1,u_2),\ldots,$ $U_M=[u_{M-1},1]$. In the ℓth bin the regressogram is defined as

$$\hat{f}_n(x) = \sum_{i=1}^{n} Y_i I\{X_i \in U_\ell\} \bigg/ \sum_{i=1}^{n} I\{X_i \in U_\ell\}, \quad x \in U_\ell, \ \ell=1,\ldots,M.$$

Clearly, the regressogram is a linear estimator with piecewise-constant weights

$$W_{ni}(x) = I\{X_i \in U_\ell\} \bigg/ \sum_{i=1}^{n} I\{X_i \in U_\ell\}, \quad x \in U_\ell, \ \ell=1,\ldots,M.$$

These weights are similar to the weights of the Nadaraya-Watson kernel estimator defined in Section 1.3. The difference is that the indicators $I\{X_i \in U_\ell\}$ appear instead of the Nadaraya-Watson kernels $K((X_i-x)/h_n)$.

Proceeding as in Section 1.4 we note that the regressogram is the solution of minimization problem

$$\min_{\theta \in R^1} \sum_{i=1}^{n} (Y_i-\theta)^2 I\{X_i \in U_\ell\}.$$

Its generalization (which is polynomial of order k in each bin) can be expressed as

$$\hat{f}_n(x) = \langle\tilde{\theta}_{n\ell}, \mathbf{F}((x-u_{\ell-1})/\delta_n)\rangle, \quad x \in U_\ell, \tag{1.13}$$

where $\tilde{\theta}_{n\ell}$ is a solution of the following minimization problem

$$(1.14) \qquad \tilde{\theta}_{n\ell} = \underset{\theta \in R^{k+1}}{\operatorname{argmin}} \sum_{i=1}^{n} (Y_i - \langle \theta, \mathbf{F}((X_i - u_{\ell-1})/\delta_n)\rangle)^2 I\{X_i \in U_\ell\}.$$

Note that the components of $\tilde{\theta}_{n\ell}$ are the estimators of coefficients in Taylor expansions of f around $u_{\ell-1}$'s, the left endpoints of bins, while in Section 1.4 for the LPE we used the expansions around a current point x.

1.5.1. DEFINITION. The <u>piecewise-polynomial estimator of order</u> k (or PPE(k)) is the function $\hat{f}_n(x)$ defined by (1.13).

The value δ_n is called <u>bin width</u>. As the bandwidth h_n of locally-polynomial estimators it controls the amount of smoothness: large values of δ_n result in higher bias, and small values lead to higher variance. For the minimization problem in (1.14) to be non-degenerate it is necessary to require that the number of points X_i in each bin is larger than $k+1$.

The solution $\tilde{\theta}_{n\ell}$ satisfies the system of normal equations (cf. (1.11))

$$\mathcal{B}_\ell \tilde{\theta}_{n\ell} = \mathbf{a}_\ell$$

where

$$\mathcal{B}_\ell = (n\delta_n)^{-1} \sum_{i=1}^{n} \mathbf{F}((X_i - u_{\ell-1})/\delta_n)\mathbf{F}^T((X_i - u_{\ell-1})/\delta_n) I\{X_i \in U_\ell\}$$

and

$$\mathbf{a}_\ell = (n\delta_n)^{-1} \sum_{i=1}^{n} Y_i \mathbf{F}((X_i - u_{\ell-1})/\delta_n) I\{X_i \in U_\ell\}.$$

In the following we impose such conditions on X_i that the matrices $\mathcal{B}_\ell$ are nonsingular for n large enough. Thus,

$$(1.15) \qquad \tilde{\theta}_{n\ell} = \mathcal{B}_\ell^{-1} \mathbf{a}_\ell.$$

It is easily seen that the piecewise-polynomial estimator is a linear one, i.e.

$$\hat{f}_n(x) = \sum_{i=1}^{n} Y_i W_{ni}^{**}(x)$$

where, for $x \in U_\ell$, the weights are defined as

$$W^{**}_{ni}(x) = \langle \mathbf{F}((x-u_{\ell-1})/\delta_n), (n\delta_n)^{-1}\mathcal{B}_\ell^{-1}\mathbf{F}((X_i-u_{\ell-1})/\delta_n)I\{X_i \in U_\ell\}\rangle.$$

The properties of piecewise-polynomial estimator are similar to those of the locally-polynomial estimator:

1.5.2. PROPOSITION.

(i) The PPE(0) is piecewise-constant estimator (the regressogram).

(ii) Assume that the errors $\xi_i \equiv 0$ (resp. $Y_i = f(X_i)$), f is a polynomial of order less or equal to k, $\hat{f}_n$ is PPE(k) and the matrices $\mathcal{B}_\ell$, $\ell=1,\dots,M$, are nonsingular. Then for all $x \in [0,1]$

$$\hat{f}_n(x) = f(x) \quad (a.s.),$$

or, equivalently,

$$\sum_{i=1}^{n} f(X_i)W^{**}_{ni}(x) = f(x) \quad (a.s.).$$

(iii) Assume that f is a polynomial of order less or equal to k, $\hat{f}_n$ is PPE(k) and the matrices $\mathcal{B}_\ell$, $\ell=1,\dots,M$, are nonsingular. If the design is deterministic then $\hat{f}_n$ is unbiased:

$$E_f(\hat{f}_n(x)) = f(x), \ \forall\ x \in [0,1].$$

1.6. BIAS AND VARIANCE OF THE ESTIMATORS

As we mentioned already, the choice of smoothing parameters h_n, δ_n controls the bias and variance of piecewise-polynomial and locally-polynomial estimators. There exists a certain dependence between bias and variance: the higher the bias the smaller the variance, and vice versa. Here we discuss this effect in detail. We obtain the explicit rates of convergence for bias and variance.

Let $\hat{f}_n$ be some estimator of a regression function f. The bias and variance of $\hat{f}_n(x)$ at a fixed point x are defined as

$$b(x) = E_f(\hat{f}_n(x)) - f(x),$$

$$\sigma^2(x) = E_f\Big((\hat{f}_n(x) - E_f(\hat{f}_n(x)))^2\Big)$$

respectively. The mean squared error (MSE) of the estimator $\hat{f}_n$ at point x equals to the squared bias plus variance:

$$\text{MSE} = \text{MSE}(\hat{f}_n, f, x) = E_f\Big((\hat{f}_n(x) - f(x))^2 \Big) = b^2(x) + \sigma^2(x).$$

The mean integrated squared error (MISE) of $\hat{f}_n$,or L_2-error, is

$$\text{MISE}=\text{MISE}(\hat{f}_n, f) = E_f\Big(\int (\hat{f}_n(x) - f(x))^2 dx\Big) = \int (b^2(x) + \sigma^2(x))dx.$$

Let us calculate the bias b(x) and the variance $\sigma^2(x)$ for the piecewise-polynomial estimator. First, consider the case of deterministic design. With random design the bias - variance calculations are more involved in view of the "random denominator" present in (1.12). The case of random design will be discussed later in this section. Proceeding to the deterministic design we assume, moreover, that it is the regular grid design. Thus, the observations have the form

$$Y_i = f(i/n) + \xi_i, \quad i=1, \ldots ,n,$$

where ξ_i are i.i.d. random variables.

1.6.1. THEOREM. Let $\hat{f}_n$ be PPE(k), $k = \lfloor\beta\rfloor$, with bin width δ_n such that $\delta_n \to 0$, $n\delta_n \to \infty$ as $n\to\infty$. Let $X_i = i/n$, $i=1,\ldots,n$, and $E(\xi_i)=0$, $E(\xi_i^2) \le \sigma^2_{max} < \infty$. Then uniformly over $f\in\Sigma(\beta,L)$, $x\in[0,1]$,

$$\text{(1.16)} \qquad b^2(x) = O(\delta_n^{2\beta}), \quad n\to\infty,$$

$$\text{(1.17)} \qquad \sigma^2(x) = O(\sigma^2_{max}(n\delta_n)^{-1}), \quad n\to\infty.$$

To prove the theorem use the following two lemmas.

1.6.2. LEMMA. Let $\phi(u)\ge 0$ be a function on R^1 which is positive on a set of positive Lebesgue measure. Then the matrix

$$\mathcal{B}^* = \int \mathbf{F}(u)\mathbf{F}^T(u)\phi(u)du$$

is positive definite.

Proof of the lemma. The integral

$$\mathbf{v}^T \mathcal{B}^* \mathbf{v} = \int \langle \mathbf{v}, \mathbf{F}(u) \rangle^2 \phi(u)\,du$$

is positive for any $\mathbf{v} \in R^{k+1}$, $\mathbf{v} \neq 0$, since the polynomial

$$\bar{\phi}(u) = \langle \mathbf{v}, \mathbf{F}(u) \rangle^2$$

has only finite number of zeros, and otherwise is strictly positive.

1.6.3. LEMMA. There exists a constant $n_0 > 0$ that does not depend on ℓ and a constant $c > 0$ that does not depend on ℓ, n such that the matrices $\mathcal{B}_\ell$ defined in Section 1.5 are positive definite for $n \geq n_0$, and

(1.18) $$|\mathcal{B}_\ell^{-1} \mathbf{v}| \leq c|\mathbf{v}|$$

for $\mathbf{v} \in R^{k+1}$, $n \geq n_0$, $\ell = 1, \ldots, M$. Here and later $|\mathbf{v}|$ denotes the Euclidean norm of a vector $\mathbf{v}$.

Proof of the lemma. Let ℓ be fixed. Denote

$$z_i = z_{i\ell} = (X_i - u_{\ell-1})/\delta_n.$$

Then

$$\mathcal{B}_\ell = (n\delta_n)^{-1} \sum_{i=1}^{n} \mathbf{F}(z_i) \mathbf{F}^T(z_i) \phi(z_i)$$

where $\phi(z) = I\{0 \leq z < 1\}$ for $\ell = 1, \ldots, M-1$, and $\phi(z) = I\{0 \leq z \leq 1\}$ for $\ell = M$. Since $\{X_i\}$ form a regular grid on $[0,1]$ with step $1/n$ we have that $\{z_{i\ell}\}$ also form a regular grid on $[0,1]$ with step $(n\delta_n)^{-1}$ (but this grid has a shift which depends on ℓ). Hence for any $\mathbf{v} \in R^{k+1}$

$$\langle \mathbf{v}, \mathcal{B}_\ell \mathbf{v} \rangle \to \langle \mathbf{v}, \mathcal{B}^* \mathbf{v} \rangle,$$

as $n \to \infty$, by the summation theorem for Riemann integrals. Here

$$\mathcal{B}^* = \int_0^1 \mathbf{F}(u) \mathbf{F}^T(u)\,du$$

Together with Lemma 1.6.2 this implies that the matrix $\mathcal{B}_\ell$ is positive definite for n large enough. Indeed, the minimal eigenvalue of $\mathcal{B}_\ell$ is greater than $\lambda_0/2$ for n large enough where $\lambda_0>0$ is the minimal eigenvalue of $\mathcal{B}^*$. Thus, the eigenvalues of the inverse matrix $\mathcal{B}_\ell^{-1}$ are uniformly bounded for n large, and (1.18) follows for any fixed ℓ.

To show that c does not depend on ℓ assume that the points u_ℓ are on the regular grid, i.e. for each ℓ there exists $j\in\{1,\dots,n\}$ such that $u_\ell=j/n$. With this assumption and the assumption of regular grid design ($X_i=i/n$, $i=1,\dots,n$) we have

$$\mathcal{B}_1 = \dots = \mathcal{B}_{M-1} = \mathcal{B}_M - (n\delta_n)^{-1}\mathbf{F}(1)\mathbf{F}^T(1)$$

The correction term for $\mathcal{B}_M$ appears since U_M is defined as a closed interval containing the right endpoint. Since $n\delta_n\to\infty$ this term is negligible, and thus (1.18) is uniform in ℓ. A similar argument is applied if the u_ℓ's are not on the regular grid: in this case all $\mathcal{B}_\ell$ may contain asymptotically negligible additive terms.

<u>Proof of Theorem 1.6.1</u>. Write

$$Y_i = p_{u_{\ell-1}}(X_i) + (f(X_i) - p_{u_{\ell-1}}(X_i)) + \xi_i$$

where $p_u(X_i)$ is the Taylor polynomial for f around u:

$$p_u(X_i) = f(u) + f'(u)(X_i-u) + \dots + f^{(k)}(u)(X_i-u)^k/k!.$$

Let n_0 be the constant from Lemma 1.6.3. Assume that $n\ge n_0$. By Lemma 1.6.3 the matrices $\mathcal{B}_\ell$ are nonsingular, hence we can apply Proposition 1.5.2 (ii). This gives

$$\text{(1.19)}\qquad E_f(\hat{f}_n(x)) = p_{u_{\ell-1}}(x) + E_f(\Delta^{**}(x)) = p_{u_{\ell-1}}(x) + \Delta^{**}(x)$$

where

$$\Delta^{**}(x) = \sum_{i=1}^{n} W_{ni}^{**}(x)(f(X_i) - p_{u_{\ell-1}}(X_i)),\ x\in U_\ell,\ \ell=1,\dots,M.$$

Expanding f in Taylor series around $u_{\ell-1}$ we get

$$f(X_i) = f(u_{\ell-1}) + f'(u_{\ell-1})(X_i-u_{\ell-1}) + \ldots$$

$$\ldots + f^{(k-1)}(u_{\ell-1})(X_i-u_{\ell-1})^{k-1}/(k-1)! +$$

$$+ \frac{(X_i-u_{\ell-1})^k}{(k-1)!}\int_0^1 f^{(k)}(u_{\ell-1}+t(X_i-u_{\ell-1}))(1-t)^{k-1}dt.$$

Since $f\in\Sigma(\beta,L)$ then for $X_i\in U_\ell$

(1.20) $$|f(X_i) - p_{u_{\ell-1}}(X_i)| \le$$

$$\le \frac{|X_i-u_{\ell-1}|^k}{(k-1)!}\int_0^1 |f^{(k)}(u_{\ell-1}+t(X_i-u_{\ell-1})) - f^{(k)}(u_{\ell-1})|\ (1-t)^{k-1}dt \le$$

$$\le L|X_i-u_{\ell-1}|^\beta/k! \le L\delta_n^\beta/k!.$$

Recall that $W_{ni}^{**}(x)$, $x\in U_\ell$, is nonzero iff $X_i\in U_\ell$. Hence, for $x\in U_\ell$,

(1.21) $$|\Delta^{**}(x)| \le (L\delta_n^\beta/k!)\sum_{i=1}^n |W_{ni}^{**}(x)|.$$

Now, using (1.18) and the definition of W_{ni}^{**} we get

(1.22) $$|W_{ni}^{**}(x)| \le c_1|\mathcal{B}_\ell^{-1}\mathbf{v}_i| \le c_2|\mathbf{v}_i|$$

where

$$\mathbf{v}_i = (n\delta_n)^{-1}\mathbf{F}(z_i)I\{X_i\in U_\ell\},$$

$$|\mathbf{v}_i| = (n\delta_n)^{-1}|\mathbf{F}(z_i)I\{X_i\in U_\ell\}| \le (n\delta_n)^{-1}|\mathbf{F}(z_i)|I\{0\le z_i\le 1\}.$$

In (1.22) and later we denote by c_i positive constants.

The sum

$$(n\delta_n)^{-1}\sum_{i=1}^{n}|\mathbf{F}(z_i)|I\{0\le z_i\le 1\}$$

tends to

$$\int_0^1 |\mathbf{F}(z)|dz < \infty ,$$

as $n\to\infty$, by the Riemann summation theorem. This together with (1.22) entails that

$$\text{(1.23)}\qquad \max_x |W_{ni}^{**}(x)| \le c_3(n\delta_n)^{-1}, \quad \sum_{i=1}^{n}|W_{ni}^{**}(x)| \le c_4 .$$

It follows from (1.21), (1.23) that

$$|\Delta^{**}(x)| \le (c_4 L/k!)\delta_n^{\beta} .$$

Also $|f(x) - p_{u_{\ell-1}}(x)| \le L\delta_n^{\beta}/k!$ for $x\in U_\ell$ (see (1.20). Using these inequalities and (1.19) we obtain (1.16).

Let us estimate the variance $\sigma^2(x)$. By the i.i.d. property of ξ_i and (1.23) we find that for $x\in U_\ell$

$$\sigma^2(x) = E_f\left((\hat{f}_n(x) - E_f(\hat{f}_n(x)))^2\right) = E\left(\left(\sum_{i=1}^{n} W_{ni}^{**}(x)\xi_i\right)^2\right) \le$$

$$\le \sigma_{max}^2 \sum_{i=1}^{n}(W_{ni}^{**}(x))^2 \le$$

$$\le \sigma_{max}^2 c_3(n\delta_n)^{-1}\sum_{i=1}^{n}|W_{ni}^{**}(x)| \le \sigma_{max}^2 c_3 c_4(n\delta_n)^{-1},$$

uniformly in ℓ,n,x. This proves (1.17). □

With minor changes the techniques of Theorem 1.6.1 can be applied to evaluate the bias and variance of locally-polynomial estimators. Introduce the following assumption on the kernel.

1.6.4. ASSUMPTION. The kernel K is bounded, nonnegative, compactly supported, and satisfies (1.4); the values of K are greater than a positive constant in some nonempty interval.

To simplify the notation we assume in the following that K is strictly positive in [-1,1], and thus

$$(1.24)\qquad K(u) \ge \kappa_0 I\{|u| \le 1\}$$

for some $\kappa_0>0$. All practical examples of kernels are symmetric functions, and they satisfy (1.24) under appropriate rescaling. Also Assumption 1.6.4 implies

$$(1.25)\qquad K(u) \le \kappa_1 I\{|u| \le \kappa_1\}$$

for some $\kappa_1>0$ since K is bounded and compactly supported.

1.6.5. THEOREM. Let $\hat{f}_n$ be LPE(k), $k = \lfloor\beta\rfloor$, with bandwidth h_n such that $h_n\to 0$, $nh_n\to\infty$ as $n\to\infty$. Let $X_i=i/n$, $i=1,\dots,n$, and $E(\xi_i)=0$, $E(\xi_i^2)\le\sigma^2_{max}<\infty$. If Assumption 1.6.4 holds then uniformly over $f\in\Sigma(\beta,L)$, $x\in[0,1]$,

$$(1.26)\qquad b^2(x) = \mathcal{O}(h_n^{2\beta}),\ n\to\infty,$$

$$(1.27)\qquad \sigma^2(x) = \mathcal{O}(\sigma^2_{max}(nh_n)^{-1}),\ n\to\infty.$$

<u>Proof</u>. Let $x\in[0,1]$ be fixed. Then for n large enough we have

$$(1.28)\qquad |\mathcal{B}^{-1}\mathbf{v}| \le c_5|\mathbf{v}|$$

for $\mathbf{v}\in R^{k+1}$ where $c_5>0$ does not depend on n and x. This is proved as in Lemma 1.6.3. In fact, (1.24) implies that $K(u)\ge\kappa_0 I\{0\le u\le 1\}$, and thus for any $\mathbf{v}\in R^{k+1}$

$$\langle\mathbf{v},\mathcal{B}\mathbf{v}\rangle = (nh_n)^{-1}\sum_{i=1}^{n}\langle\mathbf{v},\mathbf{F}(z_i)\rangle^2 K(z_i) \ge$$

$$\ge \kappa_0(nh_n)^{-1}\sum_{i=1}^{n}\langle\mathbf{v},\mathbf{F}(z_i)\rangle^2 I\{0\le z_i\le 1\}$$

where $z_i=(X_i-x)/h_n$. If $x<1-h_n$ the last expression up to the constant factor κ_0 coincides with $\langle\mathbf{v},\mathcal{B}_\ell\mathbf{v}\rangle$ (if one changes the notation as $x=u_{\ell-1}$, $h_n=\delta_n$), and by the Riemann summation theorem

$$(1.29) \qquad (nh_n)^{-1} \sum_{i=1}^{n} \langle \mathbf{v}, \mathbf{F}(z_i)\rangle^2 I\{0 \le z_i \le 1\} \to \int_0^1 \langle \mathbf{v}, \mathbf{F}(z)\rangle^2 dz$$

as $n \to \infty$. By the same argument as in Lemma 1.6.3 one gets (1.28). If $x \in [1-h_n, 1]$ the proof is similar (the complementary inequality for K is used: $K(u) \ge \kappa_0 I\{-1 \le u \le 0\}$).

By Proposition 1.4.2(ii) and the identity $p_x(x) = f(x)$ we find similarly to (1.19) that for locally-polynomial estimator

$$(1.30) \qquad E_f(\hat{f}_n(x)) = f(x) + E_f(\Delta^*(x)) = f(x) + \Delta^*(x)$$

where

$$\Delta^*(x) = \sum_{i=1}^{n} W_{ni}^*(x)(f(X_i) - p_x(X_i)).$$

By definition of W_{ni}^* and (1.28)

$$(1.31) \qquad |W_{ni}^*(x)| \le c_6 |\mathbf{v}_i|$$

where $\mathbf{v}_i = (nh_n)^{-1}\mathbf{F}(z_i)K(z_i)$. It follows from (1.25) that

$$(1.32) \qquad |W_{ni}^*(x)| \le c_6 \kappa_1 (nh_n)^{-1} |\mathbf{F}(z_i)| I\{-\kappa_1 \le z_i \le \kappa_1\}.$$

By the Riemann summation theorem

$$(1.33) \qquad (nh_n)^{-1} \sum_{i=1}^{n} |\mathbf{F}(z_i)| I\{-\kappa_1 \le z_i \le \kappa_1\} \to \int_{-\kappa_1}^{\kappa_1} |\mathbf{F}(z)| dz, \quad n \to \infty.$$

Using the same argument as after the formula (1.22) one gets the following analogue to (1.23):

$$(1.34) \qquad \max_x |W_{ni}^*(x)| \le c_7 (nh_n)^{-1}, \qquad \sum_{i=1}^{n} |W_{ni}^*(x)| \le c_8,$$

and thus

$$|\Delta^*(x)| \le c_9 h_n^{\beta}.$$

In view of (1.30) this entails (1.26).

The evaluation of the variance $\sigma^2(x)$ is straightforward with the use of (1.34):

$$\sigma^2(x) = E\left(\left(\sum_{i=1}^{n} W_{ni}^*(x)\xi_i\right)^2\right) \le \sigma_{max}^2 \sum_{i=1}^{n} (W_{ni}^*(x))^2 \le \sigma_{max}^2 c_7 c_8 (nh_n)^{-1}.$$

□

Now consider the bias - variance problem for the random design case. Denote by $b(x|\mathcal{X})$, $\sigma^2(x|\mathcal{X})$ the conditional bias and variance calculated for fixed design $\mathcal{X} = (X_1,\dots,X_n)$, i.e.

$$b(x|\mathcal{X}) = E_f(\hat{f}_n(x)|\mathcal{X}) - f(x),$$

$$\sigma^2(x|\mathcal{X}) = E_f\left((\hat{f}_n(x) - E_f(\hat{f}_n(x)|\mathcal{X}))^2|\mathcal{X}\right).$$

1.6.6. THEOREM. Let $\hat{f}_n$ be LPE(k), $k = \lfloor\beta\rfloor$, with bandwidth h_n such that $h_n \to 0$, $nh_n \to \infty$ as $n \to \infty$. Let the design points X_i be i.i.d. uniformly distributed in the interval [0,1], and $E(\xi_i)=0$, $E(\xi_i^2) \le \sigma_{max}^2 < \infty$. If Assumption 1.6.4 holds then there exists a set Q_n of design vectors $\mathcal{X}$ such that

$$P(\mathcal{X} \in Q_n) \to 1, \ n \to \infty,$$

and uniformly over $f \in \Sigma(\beta, L)$, $x \in [0,1]$, the conditional bias and variance satisfy

$$b^2(x|\mathcal{X}) \le c_{10} h_n^{2\beta}$$

$$\sigma^2(x|\mathcal{X}) \le c_{11}(nh_n)^{-1}$$

for all $\mathcal{X} \in Q_n$.

Proof. In the proof of Theorem 1.6.5 the properties of regular design were used only to show the convergence in (1.29) and (1.33) (the summation formula for Riemann integrals was applied). Now, if X_i are i.i.d uniformly distributed in [0,1] then z_i are i.i.d. uniformly distributed in the interval

$[-x/h_n, (1-x)/h_n]$ where they have a constant density h_n. Using the law of large numbers it is easy to verify that (1.29), (1.33) hold in the sense of convergence in probability. Hence there exist some $c_7=c_7^*$, $c_8=c_8^*$ such that the probability that (1.34) is true converges to 1 as $n\to\infty$ (one can easily verify that the convergence is uniform in $x\in[0,1]$). In other words, if we denote by Q_n the set of designs $\mathcal{X}=(X_1,\dots,X_n)$ for which (1.34) is true with $c_7=c_7^*$, $c_8=c_8^*$ then

$$P(\mathcal{X}\in Q_n) \to 1, \quad n\to\infty,$$

uniformly in x. But if $\mathcal{X}\in Q_n$ then we can use the argument of Theorem 1.6.5 and we find that there exist c_{10}, $c_{11}>0$, independent of $n,x,\mathcal{X}$, such that the statement of the theorem holds. □

1.6.7. REMARK. Theorem 1.6.6 is also valid under more general assumptions on the design. It is sufficient to assume that X_i's are i.i.d. random variables with a continuous density $\mu(x)$ which is bounded away from 0 uniformly on [0,1]. Modification of the proof for this case is proposed as an exercise for the reader.

The theorems of this section have some important corollaries. From Theorem 1.6.1 we obtain for piecewise -polynomial estimator:

$$\sup_x \mathrm{MSE}(x) = \mathcal{O}(\delta_n^{2\beta} + (n\delta_n)^{-1}), \quad n\to\infty,$$

$$\mathrm{MISE} = \mathcal{O}(\delta_n^{2\beta} + (n\delta_n)^{-1}), \quad n\to\infty.$$

This immediately entails that the best convergence rate of MSE and MISE for piecewise-polynomial estimators is achieved with the bin width choice

$$\delta_n \sim n^{-1/(2\beta+1)}.$$

This value of bin width is called <u>optimal</u>. For optimal δ_n

$$\text{(1.35)} \qquad \begin{aligned} \sup_x \mathrm{MSE}(x) &= \mathcal{O}(n^{-2\beta/(2\beta+1)}), \quad n\to\infty, \\ \mathrm{MISE} &= \mathcal{O}(n^{-2\beta/(2\beta+1)}), \quad n\to\infty. \end{aligned}$$

Similarly, Theorem 1.6.5 implies that for the

locally-polynomial estimator

$$\sup_x \mathrm{MSE}(x) = O(h_n^{2\beta} + (nh_n)^{-1}), \quad n\to\infty,$$

$$\mathrm{MISE} = O(h_n^{2\beta} + (nh_n)^{-1}), \quad n\to\infty.$$

The bandwidth which minimizes MSE and MISE asymptotically is

$$h_n \sim n^{-1/(2\beta+1)}.$$

This value is called <u>optimal</u> bandwidth. For the optimal h_n the MSE and MISE of locally-polynomial estimator satisfy (1.35).

The optimal choice of δ_n and h_n provides the balance between bias and variance contributions to the asymptotic error: the squared bias and the variance have the same order under this choice.

We summarize these remarks in the following theorem.

1.6.8. THEOREM. If $\hat{f}_n$ is PPE(k) and assumptions of Theorem 1.6.1 are satisfied with $\delta_n=\tilde{c}n^{-1/(2\beta+1)}$, or $\hat{f}_n$ is LPE(k) and assumptions of Theorem 1.6.5 are satisfied with $h_n=\tilde{c}n^{-1/(2\beta+1)}$, $\tilde{c}>0$, then, as $n\to\infty$,

$$\sup_{f\in\Sigma(\beta,L)} \sup_x E_f\left((\hat{f}_n(x) - f(x))^2\right) = O(n^{-2\beta/(2\beta+1)}),$$

$$\sup_{f\in\Sigma(\beta,L)} E_f\left(\int(\hat{f}_n(x) - f(x))^2 dx\right) = O(n^{-2\beta/(2\beta+1)}).$$

For the random design case similar results are not true under the assumptions of Theorem 1.6.6 (although, they can be proved under much more restrictive assumptions). Theorem 1.6.6 implies only a weaker result, with convergence in probability instead of mean squared convergence.

1.6.9. THEOREM. If $\hat{f}_n$ is LPE(k) and assumptions of Theorem 1.6.6 are satisfied with $h_n=\tilde{c}n^{-1/(2\beta+1)}$, $\tilde{c}>0$, then, as $C\to\infty$,

$$\limsup_{n\to\infty} \sup_{f\in\Sigma(\beta,L)} \sup_x P_f\left(n^{\beta/(2\beta+1)}|\hat{f}_n(x) - f(x)| \geq C\right) = \tag{1.36}$$

$$= O(C^{-2}),$$

$$\limsup_{n\to\infty} \sup_{f\in\Sigma(\beta,L)} P_f\left(n^{\beta/(2\beta+1)}\left(\int(\hat{f}_n(x) - f(x))^2dx\right)^{1/2} \geq C\right) = \tag{1.37}$$

$$= O(C^{-2}).$$

Proof. By Chebyshev's inequality

$$P_f\left(n^{\beta/(2\beta+1)}|\hat{f}_n(x) - f(x)| \geq C\right) \leq$$

$$\leq P_f\left(n^{2\beta/(2\beta+1)}(\hat{f}_n(x) - f(x))^2 \geq C^2,\ \mathcal{X}\in Q_n\right) + P(\mathcal{X}\notin Q_n) \leq$$

$$\leq n^{2\beta/(2\beta+1)}E\left(\text{MSE}(x|\mathcal{X})I\{\mathcal{X}\in Q_n\}\right)C^{-2} + P(\mathcal{X}\notin Q_n).$$

where Q_n is as in Theorem 1.6.6. By definition of Q_n (see the proof of Theorem 1.6.6) and definition of h_n

$$\text{MSE}(x|\mathcal{X}) = b^2(x|\mathcal{X}) + \sigma^2(x|\mathcal{X}) \leq c_{10}^2h_n^{2\beta} + c_{11}(nh_n)^{-1} =$$

$$= (c_{10}^2\tilde{c}^{2\beta} + c_{11}\tilde{c}^{-1})n^{-2\beta/(2\beta+1)},\quad \mathcal{X}\in Q_n,$$

uniformly over $f\in\Sigma(\beta,L)$, $x\in[0,1]$. This together with the uniform in x convergence

$$P(\mathcal{X}\in Q_n) \to 1,\ n\to\infty,$$

implies (1.36); (1.37) is proved similarly. □

1.7. CRITERIA FOR COMPARING THE NONPARAMETRIC ESTIMATORS

The MSE and MISE error criteria were introduced as measures of efficiency of nonparametric regression estimators for a particular function f. These criteria are examples of risk functions. In general, a regression function f is considered as an element of some linear space equipped with a pseudometric $d(\cdot,\cdot)$. Remind that pseudometric is defined as a function, satisfying the same conditions as metric, except, possibly, the condition: "$d(f,g) = 0 \Rightarrow f = g$". The following pseudometrics are commonly used in nonparametric estimation problems:

$$d(f,g) = \int |f-g| \qquad (L_1\text{-metric}),$$

$$d(f,g) = \left(\int (f-g)^2\right)^{1/2} \qquad (L_2\text{-metric}),$$

$$d(f,g) = |f(x_0)-g(x_0)| \qquad (\text{distance at a fixed point } x_0),$$

$$d(f,g) = \sup_x |f(x)-g(x)| \qquad (\text{uniform metric, or C-metric}).$$

The risk function is defined as

$$R(\hat{f}_n, f) = E_f\left(w(\psi_n^{-1} d(\hat{f}_n, f)\right)$$

where w is a loss function, and $\psi_n \to 0$ is a normalizing positive sequence which is interpreted as the rate of convergence.

The loss function $w(\cdot)$ is assumed to be a nonnegative real-valued function. The following examples of loss functions will be used in this book: $w(u) = u^2$ (squared loss), $w(u) = |u|$ (absolute value loss) and $w(u)=I\{|u|\ge C\}$ where $C>0$ is a constant (indicator loss).

1.7.1. EXAMPLES. (i) Squared loss function, $w(u)=u^2$.

If $d(\cdot,\cdot)$ is the L_2-metric one gets

$$R(\hat{f}_n, f) = \psi_n^{-2} \mathrm{MISE}(\hat{f}_n, f).$$

If $d(\cdot,\cdot)$ is the distance at a fixed point x_0 then

$$R(\hat{f}_n, f) = \psi_n^{-2} \mathrm{MSE}(\hat{f}_n, f, x_0).$$

It is natural to choose normalizing sequences ψ_n in such a way that the risk $R(\hat{f}_n, f)$ remains separated from 0 and ∞ as n increases. Thus, if $\hat{f}_n$ is a piecewise-polynomial estimator and $f\in\Sigma(\beta, L)$ one takes $\psi_n = n^{-\beta/(2\beta+1)}$ (cf. (1.23)).

(ii) Indicator loss function. The risk is a probability that the normalized deviation of $\hat{f}_n$ exceeds a given threshold $C>0$:

$$R(\hat{f}_n, f) = P_f\left(\psi_n^{-1} d(\hat{f}_n, f) \geq C\right).$$

Risk functions are used to compare different estimators. For each fixed estimator they are functions of f. To compare the risks $R(\hat{f}_{n1}, f)$ and $R(\hat{f}_{n2}, f)$ for some estimators $\hat{f}_{n1}$, $\hat{f}_{n2}$ one has to order them. Unfortunately, the attempts to order risk functions for all f fail. For example, consider the risk function

$$R(\hat{f}_n, f) = n^{2\beta/(2\beta+1)} \mathrm{MSE}(\hat{f}_n, f, x_0). \tag{1.38}$$

If $\hat{f}_{n1}$ is the locally-polynomial estimator of order $k = \lfloor\beta\rfloor$ with bandwidth $h_n = n^{-1/(2\beta+1)}$ and X_i form the regular design then, as follows from Theorem 1.6.5, $R(\hat{f}_{n1}, f)$ is bounded as $n\to\infty$ uniformly over $f\in\Sigma(\beta, L)$. Now, consider the absurd estimator $\hat{f}_{n2} = \hat{f}_{n2}(x) \equiv 0$. For any fixed regression function f such that $f(x_0)>0$ the risk $R(\hat{f}_{n2}, f)$ tends to ∞ ,as $n\to\infty$, with the rate $n^{2\beta/(2\beta+1)}$, and hence $\hat{f}_{n2}$ is worse than $\hat{f}_{n1}$ with respect to the risk (1.38). But for $f(x)\equiv 0$ the risk $R(\hat{f}_{n2}, 0)$ is equal to 0 independently of n, i.e. the estimator $\hat{f}_{n2}$ is better than $\hat{f}_{n1}$. This example shows that the difficulty in ordering the risks of type (1.38) comes from their dependence on the unknown regression function f.

This suggests to find such scalar characteristics of risks that do not depend on a particular function f, and then to order these characteristics. The scalar characteristics should be functions of the estimator $\hat{f}_n$ only.

One can propose many examples of scalar characteristics. Although, there exists a tradition in statistics (see e.g. Ibragimov and Khasminskii (1981)) to consider only two of them: the maximal risk and the Bayesian risk.

Let $R(\hat{f}_n, f)$ be a risk function and let Σ be a set of functions that contains the "true" regression function f.

1.7.2. DEFINITION. The value

$$r(\hat{f}_n) = \sup_{f\in\Sigma} R(\hat{f}_n, f)$$

is called maximal risk of the estimator $\hat{f}_n$ on the set Σ.

If it is possible to define a probability measure $\Omega_n(df)$ on the set Σ then one can also introduce the Bayesian risk.

1.7.3. DEFINITION. The value

$$r^B(\hat{f}_n) = \int_\Sigma R(\hat{f}_n, f)\Omega_n(df)$$

is called Bayesian risk of the estimator $\hat{f}_n$ on the set Σ.

We consider only nonnegative risk functions R, therefore both definitions are correct if one allows to assign the value $+\infty$ to r and r^B.

If Σ is a nonparametric class of functions then it is not always straightforward to define a probability measure Ω_n on it. This is the reason why the Bayesian risk is rarely used in nonparametric problems.

Assume that we have ordered nonparametric estimators according to the values of their maximal risks. The best estimator on this scale is the one for which the maximal risk attains its minimum (over all possible estimators), i.e. the estimator $\hat{f}_n$ such that

$$r(\hat{f}_n) = \min_{T_n} r(T_n) = \min_{T_n} \sup_{f\in\Sigma} R(T_n, f)$$

($\min_{T_n}$ denotes the minimum over all possible estimators). If this is true $\hat{f}_n$ is called minimax estimator. The value

$$\min_{T_n} \sup_{f\in\Sigma} R(T_n, f)$$

is called minimax risk on Σ. The construction of minimax nonparametric regression estimators for different classes Σ is a hard problem. It is solved only asymptotically for some special cases. More general and rough approach consists in comparing the estimators by the convergence rates of their maximal risks. For example, Theorem 1.6.8 shows that for piecewise-polynomial and locally-polynomial estimators the maximal squared risks on $\Sigma=\Sigma(\beta,L)$ have convergence rates $n^{-2\beta/(2\beta+1)}$. We are not interested for the moment in the value of the constant factor that multiplies the rate $n^{-2\beta/(2\beta+1)}$. The question is whether this rate can be improved by other estimators. If not, then $\hat{f}_n$ is optimal estimator (i.e. the estimator having the optimal rate of convergence). The formal definition of optimal estimator is delayed to Section 2.1 where more general statistical framework is considered. The results of Chapter 2 show that the piecewise-polynomial and locally-polynomial estimators as in Theorem 1.6.8 are in fact optimal estimators.

1.8. RATES OF THE UNIFORM AND L_1- CONVERGENCE

As shown in Section 1.6 the best convergence rates of locally-polynomial and piecewise-polynomial estimators on classes $\Sigma(\beta,L)$ are of order $n^{-\beta/(2\beta+1)}$. This was proved for L_2-risks and for the risks at a fixed point. Let us consider two other examples that we are interested in: the uniform metric and the L_1-metric. We show that for L_1-risks the situation is the same as for L_2-risks, that is the optimal LPE's and PPE's converge with the L_1-rate $n^{-\beta/(2\beta+1)}$. However, for the risks with uniform metric the rates are slightly worse: they are of the order $(n/\log n)^{-\beta/(2\beta+1)}$. Roughly speaking, the log-factor appears here due to the fact that the maximum of squares of M Gaussian random variables has the order $O(\log M)$ (the same is true for random variables with exponentially decreasing tails). To explain the effect of log-factor in detail we need some calculations. We present them here for the Gaussian case.

For the uniform metric and squared loss function the risk is given by

$$(1.39)\qquad R(\hat{f}_n,f) = E_f\left(\sup_x|\hat{f}_n(x) - f(x)|^2\right) \le$$

$$\le 2\sup_x|E_f(\hat{f}_n(x)) - f(x)|^2 + 2E_f\left(\sup_x|\hat{f}_n(x) - E_f(\hat{f}_n(x))|^2\right).$$

Assume that $\hat{f}_n$ is PPE(k), $k=\lfloor\beta\rfloor$, with bin width δ_n, and that the regular deterministic design is used. Then it follows from Theorem 1.6.1 that uniformly over $f\in\Sigma(\beta,L)$ the squared bias contribution is

$$\sup_x |E_f(\hat{f}_n(x)) - f(x)|^2 = O(\delta_n^{2\beta}),\ n\to\infty. \tag{1.40}$$

Now, the variance contribution can be represented as

$$E_f\left(\sup_x|\hat{f}_n(x) - E_f(\hat{f}_n(x))|^2\right) =$$

$$= E_f\left(\max_{\ell=1,\dots,M}\ \sup_{x\in U_\ell}|\hat{f}_n(x) - E_f(\hat{f}_n(x))|^2\right) =$$

$$= E_f\left(\max_{\ell=1,\dots,M}\ \sup_{x\in U_\ell}\left(\sum_{i=1}^n W_{ni}^{**}(x)\xi_i\right)^2\right).$$

It follows from the definition of W_{ni}^{**} (see Section 1.5) that

$$\sup_{x\in U_\ell}\left(\sum_{i=1}^n W_{ni}^{**}(x)\xi_i\right)^2 =$$

$$= \sup_{x\in U_\ell}\left(<\mathbf{F}((x-u_{\ell-1})/\delta_n),(n\delta_n)^{-1}\mathcal{B}_\ell^{-1}\sum_{i=1}^n \mathbf{F}((X_i-u_{\ell-1})/\delta_n)I\{X_i\in U_\ell\}\xi_i>\right)^2\le$$

$$\le c_{12}|(n\delta_n)^{-1}\mathcal{B}_\ell^{-1}\sum_{i=1}^n \mathbf{F}((X_i-u_{\ell-1})/\delta_n)I\{X_i\in U_\ell\}\xi_i|^2$$

for some $c_{12}>0$. Here $|\cdot|$ is the Euclidean norm of a vector. Hence

$$E_f\left(\sup_x|\hat{f}_n(x) - E_f(\hat{f}_n(x))|^2\right) \le$$

$$\le c_{12}(n\delta_n)^{-1}E_f\left(\max_{\ell=1,\dots,M}|(n\delta_n)^{-1/2}\mathcal{B}_\ell^{-1}\sum_{i=1}^n \mathbf{F}((X_i-u_{\ell-1})/\delta_n)I\{X_i\in U_\ell\}\xi_i|^2\right)$$

Since the matrix $\mathcal{B}_\ell$ is symmetric it follows from Lemma 1.6.3 that $|\mathcal{B}_\ell^{-1/2}\mathbf{v}| \le c|\mathbf{v}|$ for all $\mathbf{v}\in R^{k+1}$, uniformly in $n \ge n_0$, $\ell=1,\dots,M$. Thus

$$(1.41)\qquad E_f\left(\sup_x |\hat{f}_n(x) - E_f(\hat{f}_n(x))|^2\right) \le$$

$$\le c_{13}(n\delta_n)^{-1} E_f\left(\max_{\ell=1,\dots,M} |(n\delta_n\mathcal{B}_\ell)^{-1/2}\sum_{i=1}^n \mathbf{F}((X_i-u_{\ell-1})/\delta_n)I\{X_i\in U_\ell\}\xi_i|^2\right).$$

Assume that ξ_i are i.i.d. standard Gaussian random variables. Then

$$(n\delta_n\mathcal{B}_\ell)^{-1/2}\sum_{i=1}^n \mathbf{F}((X_i-u_{\ell-1})/\delta_n)I\{X_i\in U_\ell\}\xi_i, \quad \ell=1,\dots,M,$$

are i.i.d. Gaussian random vectors with mean 0 and identity covariance matrices. Thus, to evaluate the right-hand side of (1.41) it suffices to find an upper bound on $E\left(\max_{\ell=1,\dots,M} \tilde{\eta}_\ell^2\right)$ where $\tilde{\eta}_\ell$ are i.i.d. standard Gaussian random variables. It is easy to see that

$$E\left(\max_{\ell=1,\dots,M} \tilde{\eta}_\ell^2\right) = O(\log M)$$

(Galambos, 1978, Ch.1). Since $M=1/\delta_n$ the right-hand side of (1.41) is $O(\log(1/\delta_n)/n\delta_n)$. This together with (1.39), (1.40) implies

$$R(\hat{f}_n, f) = O(\delta_n^{2\beta}) + O(\log(1/\delta_n)/n\delta_n),\ n\to\infty.$$

Optimizing with respect to δ_n one gets

$$R(\hat{f}_n, f) = O((\log n/n)^{2\beta/(2\beta+1)}),\ n\to\infty,$$

where $\hat{f}_n$ is the estimator with bin width

$$\delta_n \sim (\log n/n)^{1/(2\beta+1)}.$$

Thus, we proved the following theorem.

1.8.1. THEOREM. Let $\hat{f}_n$ be PPE(k), $k = \lfloor\beta\rfloor$, with bin width $\delta_n=(\log n/n)^{1/(2\beta+1)}$. Assume that $X_i=i/n$, $i=1,\ldots,n$, and ξ_i are i.i.d. standard Gaussian random variables. Then, as $n\to\infty$,

$$\sup_{f\in\Sigma(\beta,L)} E_f\left(\sup_x|\hat{f}_n(x) - f(x)|^2\right) =$$

$$= O((\log n/n)^{2\beta/(2\beta+1)}),\ n\to\infty.$$

1.8.2. REMARK. To simplify the calculations we proved Theorem 1.8.1 under rather restrictive assumptions. In fact this result holds in more general case. For example, X_i's can be i.i.d. random variables satisfying assumptions of Remark 1.6.7 and ξ_i can be non-Gaussian zero mean variables with exponentially decreasing tails (see e.g. Assumption 1.9.7 in the next section).

Now, let $R(\hat{f}_n,f)$ be the L_1-risk. By the Cauchy-Schwartz inequality

$$R(\hat{f}_n,f) = E_f\left(\int_0^1|\hat{f}_n(x)-f(x)|dx\right) \le \left[E_f\left(\int_0^1(\hat{f}_n(x)-f(x))^2dx\right)\right]^{1/2} \le$$

$$\le \left[\mathrm{MISE}(\hat{f}_n,f)\right]^{1/2}.$$

Hence, from Theorem 1.6.1 we get the following.

1.8.3. THEOREM. Let $\hat{f}_n$ be PPE(k) and let the assumptions of Theorem 1.6.1 be satisfied. Then, uniformly in $f\in\Sigma(\beta,L)$,

$$E_f\left(\int|\hat{f}_n(x)-f(x)|dx\right) = O(\delta_n^\beta + \sigma_{max}(n\delta_n)^{-1/2}),\ n\to\infty.$$

Similar result for the L_1-rate of the LPE(k) is deduced from Theorem 1.6.5. As a consequence of this and of Theorem 1.8.3 we get:

1.8.4. THEOREM. If $\hat{f}_n$ is PPE(k) and assumptions of Theorem 1.6.1 are satisfied with $\delta_n=\tilde{c}n^{-1/(2\beta+1)}$, or $\hat{f}_n$ is LPE(k) and assumptions of Theorem 1.6.5 are satisfied with $h_n=\tilde{c}n^{-1/(2\beta+1)}$, $\tilde{c}>0$, then, as $n\to\infty$,

$$\sup_{f\in\Sigma(\beta,L)} E_f\left(\int|\hat{f}_n(x)-f(x)|dx\right) = O(n^{-\beta/(2\beta+1)}).$$

1.9. THE CHANGE-POINT PROBLEM

As it was mentioned in the Introduction, our approach to image reconstruction uses some one-dimensional change-point models. Here we discuss these basic models in some detail.

Let the observations Y_i be defined by

$$Y_i = I(X_i \le \vartheta) + \xi_i, \quad i = 1,\dots,n, \tag{1.42}$$

where ξ_i are i.i.d. (0,1)-Gaussian random variables; X_i are design points within the interval [0,1]; $I(\cdot)$ is the indicator function; ϑ is unknown parameter (change-point parameter) which is to be estimated from observations (1.42). It is assumed that $h<\vartheta<1-h$ where h is some known value, $0<h<1/2$.

In this section we distinguish between several types of designs:

(A1) <u>regular grid design</u>: $X_i = i/n$, $i=1,\dots,n$;

(A2) <u>random shift of the regular grid</u>: $X_i = i/n-\eta_1$, $i=1,\dots,n$, (η_1 is a random variable uniformly distributed in $[0,n^{-1}]$);

(B1) <u>random design</u> with independent X_i uniformly distributed in the interval [0,1] ;

(B2) <u>jittered regular grid</u>: X_i are independent random variables having different distributions: X_i is uniformly distributed in the interval $[(i-1)/n, i/n]$, $i=1,\dots,n$.

It is supposed that all observations and design points are available to the statistician, thus the change-point problem is studied here in a posteriori setup. Therefore, we avoid all the questions arising in sequential analysis where the observations appear successively and statistical inference is based on some Markovian stopping rules (cf. Shiryaev (1978), Siegmund (1985)).

For each estimator $\hat{\vartheta}_n = \hat{\vartheta}_n((X_1,Y_1),\dots,(X_n,Y_n))$ of the change-point parameter ϑ we introduce the squared risk

$$R(\hat{\vartheta}_n,\vartheta) = \psi_n^{-2}\, E_\vartheta\left((\hat{\vartheta}_n-\vartheta)^2\right)$$

and the maximal risk

$$r(\hat{\vartheta}_n) = \sup_{h<\vartheta<1-h} R(\hat{\vartheta}_n,\vartheta)$$

where ψ_n is a fixed sequence (the rate of convergence), and E_ϑ is the expectation with respect to the joint distribution of (X_i,Y_i), $i=1,\ldots,n$.

In this section we show that there exist the change-point estimators which converge with the rate $\psi_n=n^{-1}$, or in other words, that with this choice of ψ_n the maximal risk $r(\hat{\vartheta}_n)$ is bounded for n large. Also an asymptotic expansion for the maximum likelihood estimator of change-point parameter will be derived.

We start with some remarks concerning the "continuous-time" analogue of the model (1.42). We do this since the results related to the continuous-time change-point problem are transparent and useful to understand the plan of our exposition in general. Now we give a short overview of them following Ibragimov and Khasminskii (1981) where all the details can be found. (Readers who are not familiar with continuous-time estimation problems may skip this part of the text which we mark for convenience by signs **).

** The continuous analogue to the model (1.42) with the regular design is

(1.43) $$dY^{(n)}(t) = I(t\le\vartheta)dt + n^{-1/2}dW(t)\ ,\quad 0\le t\le 1,$$

where W(t) is some standard Wiener process. We consider (1.43) as the Ito stochastic differential equation with zero initial condition: $Y^{(n)}(0)=0$. Relations between (1.43) and (1.42) are direct. Indeed, if $i/n \le \vartheta$, then the variables

(1.44) $$n\Big(Y^{(n)}(i/n) - Y^{(n)}((i-1)/n)\Big)$$

are independent (1,1)-Gaussian , i. e. they have the unit mean and variance, while for $(i-1)/n \ge \vartheta$ they are (0,1)-Gaussian similarly to (1.42). Note that the only possible non-coincidence in (1.42) and (1.44) may occur for $i=i_0$ such that $(i_0-1)/n < \vartheta < i_0/n$.

The continuous model (1.43) has the following asymptotic properties:

(a) The log-likelihood ratio

$$Z_n(u) = \log\left(\frac{dP_{\vartheta+u/n}}{dP_\vartheta}(Y^{(n)}(\cdot))\right)$$

converges weakly in P_ϑ-probability to the stochastic process $Z_0(u) = W_0(u) - |u|/2$ as $n \to \infty$, where

$$W_0(u) = \begin{cases} W_1(u) & \text{if } u > 0 \\ W_2(-u) & \text{if } u < 0 \end{cases}$$

is the two-sided Wiener process generated by some independent standard Wiener processes $W_1(u)$ and $W_2(u)$.

(b) Let ϑ_n^* be the maximum likelihood estimator (MLE) obtained from observations (1.43). Then the limit distribution of the random variables $n(\vartheta_n^*-\vartheta)$ coincides with that for the maximizer u_0^* of the random function $Z_0(u)$.

(c) The random variables $n(\vartheta_n^*-\vartheta)$ have all the moments converging to the moments of u_0^* which are finite; the first two of them are equal respectively to

$$E[u_0^*] = 0 \quad \text{and} \quad E[(u_0^*)^2] = 26 \ .$$

(d) The MLE is not asymptotically minimax. Namely, if n is large enough, then

$$r(\vartheta_n^B)/r(\vartheta_n^*) \approx 0.75$$

where ϑ_n^B is the Bayesian estimator for an arbitrary positive continuous a priori density and the squared loss function. This Bayesian estimator is proved to be asymptotically minimax. **

Let us return to the discrete model (1.42). Since the random variables ξ_i are Gaussian the MLE coincides with the least squares estimator (LSE) which is the solution of the following minimization problem:

$$\min_{\vartheta'} \sum_{i=1}^{n} \left(Y_i - I(X_i \le \vartheta')\right)^2$$

or, equivalently,

$$\max_{\vartheta'} J(\vartheta') \tag{1.45}$$

where

$$J(\vartheta') = \sum_{i=1}^{n} I(X_i \le \vartheta')(Y_i - \tfrac{1}{2}).$$

Note that $J(\cdot)$ is a right-continuous piecewise constant function with jumps at points X_i. We assume that maximization in (1.45) is over $\vartheta' \in [X_1, X_n)$, and thus possible boundary effects are neglected. Due to the Gaussian distribution of noises there exists a unique $\hat{i}$, $1 < \hat{i} \le n$, such that any solution of (1.45) belongs to the interval $[X_{\hat{i}-1}, X_{\hat{i}})$; this property holds almost surely with respect to P_ϑ for any ϑ. It will be convenient to define the MLE ϑ_n^* as the mean value of the extreme solutions:

$$\vartheta_n^* = (X_{\hat{i}-1} + X_{\hat{i}})/2, \quad 1 < \hat{i} \le n.$$

By analogy to the property (d) of continuous-time model we conjecture that the MLE ϑ_n^* is not the best estimator, i. e. in asymptotics it is not the minimax one but only minimax to within a constant. Its rate of convergence is $\psi_n = 1/n$, as shown below. We shall see in the next chapter that this rate cannot be improved by any other estimator uniformly in $\vartheta \in (h, 1-h)$.

Define the two-sided random walk Z_k:

$$Z_0 = 0, \quad Z_k = W_k - |k|/2, \quad k = \pm 1, \pm 2 \ldots,$$

where

$$W_k = \begin{cases} \Sigma_{i=1}^{k} \bar{\xi}_i & \text{if } k > 0, \\ \Sigma_{i=k}^{-1} \bar{\xi}_i & \text{if } k < 0, \end{cases}$$

$\bar{\xi}_i$, $i = \pm 1, \pm 2 \ldots$, are $(0,1)$ - Gaussian random variables. Let $\hat{k}$ be the maximizer of Z_k : $Z_{\hat{k}} > Z_k$ for $k \ne \hat{k}$. Note that the paths Z_k go to $-\infty$ with probability 1 as $|k| \to \infty$. Hence, the point $\hat{k}$ exists almost surely. Since $\bar{\xi}_i$ are Gaussian $\hat{k}$ is almost surely unique. Introduce the probabilities

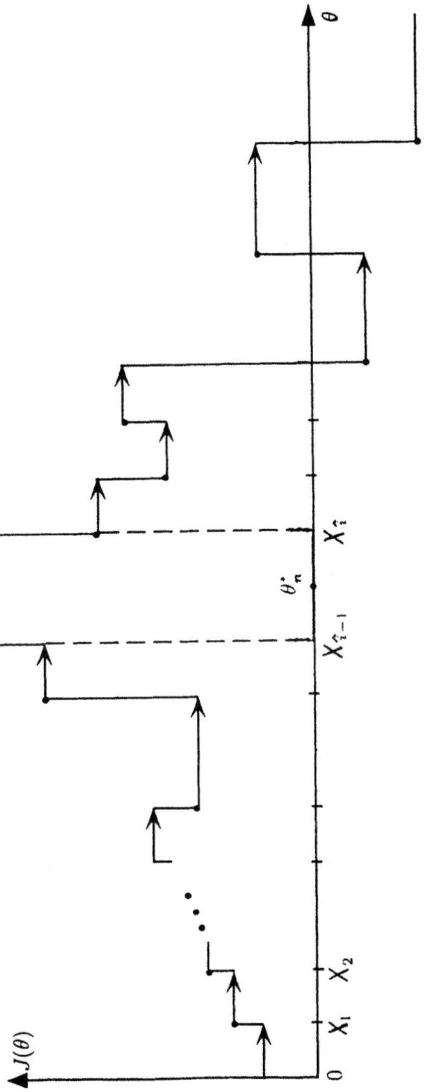

Fig. 1.1 Definition of the MLE θ_n^*.

$$\pi_k = P(\hat{k} = k), \quad k=0,\pm1,\dots .$$

1.9.1. LEMMA. The probabilities π_k are positive, symmetric, i.e. $\pi_k=\pi_{-k}$, and there exists $C_\pi>0$ such that

$$\pi_k \le \exp(-C_\pi|k|)$$

for $|k|$ large enough.

Proof. Using the well-known inequality for Mills' ratio (see e.g. Johnson and Kotz (1970), p.279) :

$$1-\Phi(x) \le (2\pi)^{-1/2}\exp(-x^2/2), \ x>1,$$

where $\Phi(x)$ is the standard Gaussian c.d.f. one gets

$$\pi_k = P(\hat{k} = k) \le P(\hat{k} \ge k) \le$$

$$\le P\left(\sup_{k_1\ge k} Z_{k_1} \ge 0\right) \le \sum_{k_1\ge k} P\left(k_1^{-1/2}W_{k_1} \ge k_1^{1/2}/2\right) =$$

$$= \sum_{k_1\ge k} (1-\Phi(k_1^{1/2}/2)) \le (2\pi)^{-1/2} \sum_{k_1\ge k} \exp(-k_1/8) \le \exp(-C_\pi k)$$

for $k>0$ large enough where C_π is some positive constant. The same argument is used for $k<0$. □

In the following we assume without loss of generality that X_i are numbered in such a way that $X_1\le X_2\le\dots\le X_n$. For the designs (A1), (A2), (B2) this ordering holds already by definition.

Let the integer $i_0=i_0(\vartheta)$ be such that $X_{i_0-1} \le \vartheta < X_{i_0}$. Note that $J(\vartheta)=J\left((X_{i_0-1}+X_{i_0})/2\right)$ by definition. The integer $\hat{i}$ which has been introduced above is the maximizer of

$$J\left((X_{i-1} + X_i)/2\right) - J\left((X_{i_0-1} + X_{i_0})/2\right)$$

over $1<i\le n$, or, equivalently, $\hat{i}-i_0$ is the maximizer of

$$J_k=J\left((X_{k+i_0-1} + X_{k+i_0})/2\right) - J\left((X_{i_0-1} + X_{i_0})/2\right)$$

over k such that $1<k+i_0\le n$. Using the definition of $J(\cdot)$ we get

$$J_k = \begin{cases} \sum_{j=i_0}^{i_0+k-1} \xi_j - |k|/2 & \text{if } k > 0, \\ -\sum_{j=i_0+k}^{i_0-1} \xi_j - |k|/2 & \text{if } k < 0, \end{cases}$$

$(1<k+i_0\le n)$. If we choose

$$\bar{\xi}_i = \begin{cases} \xi_{i+i_0-1}, & 1\le i\le n-i_0, \\ -\xi_{i+i_0}, & 1-i_0\le i\le -1, \end{cases}$$

then

$$J_k = Z_k, \ 1<k+i_0\le n.$$

Hence $\hat{i}-i_0$ is the maximizer of Z_k over k such that $1<k+i_0\le n$, and if $1<\hat{k}+i_0\le n$ then $\hat{i}=\hat{k}+i_0$. Note that the integer i_0 is random and its value depends on the design $X_1,\ldots,X_n$. Hence the only dependence of $\hat{i}$ on the design is via the set of admissible values of k, $k\in\{-i_0+2,\ldots,-i_0+n\}$. Provided this set is fixed, i. e. for the given design, the value of $\hat{i}-i_0$ is entirely defined by $\bar{\xi}_i$.

The following lemma shows that for i_0 separated from 1 and n the conditional probability

$$P_\vartheta(\hat{i}=k+i_0|X_1,\ldots,X_n), \ 1<k+i_0\le n,$$

can be approximated by π_k's.

1.9.2.LEMMA. Let a random set of design points

$$\mathfrak{A} = \{(X_1,\ldots,X_n): nh_0\le i_0\le n(1-h_0)+2\}$$

be defined for some positive h_0. Then for $n>3/h_0$ and for any design $(X_1,\ldots,X_n)\in\mathfrak{A}$ the inequalities hold

$$|P_\vartheta(\hat{i}=k+i_0|X_1,\ldots,X_n) - \pi_k| \le 2\exp(-C_\pi(nh_0-3)), \ 1<k+i_0\le n,$$

where $C_\pi>0$ is the constant from Lemma 1.9.1.

<u>Proof</u>. It follows from the definition that

$$P_\vartheta(\hat{i}=k+i_0|X_1,\ldots,X_n) =$$

$$= P_\vartheta(\hat{i}=k+i_0, \ 1<\hat{k}+i_0\le n|X_1,\ldots,X_n) + P(\hat{k}+i_0\notin\{2,\ldots,n\}) \le$$

$$\le P(\hat{k}=k,\ 1<\hat{k}+i_0\le n) + P(\hat{k}+i_0\notin\{2,\dots,n\}) \le$$

$$\le \pi_k + P(\hat{k}+i_0\notin\{2,\dots,n\}).$$

By Lemma 1.9.1

$$P(\hat{k}+i_0\notin\{2,\dots,n\}) \le \sum_{|k|\ge nh_0-2} \pi_k \le$$

$$2\sum_{k\ge nh_0-2} \exp(-C_\pi k) \le 2\exp(-C_\pi(nh_0-3)).$$

The proof of the opposite inequality is the same. □

The following proposition is true for any of the designs (A1),(A2), and (B1),(B2).

1.9.3. PROPOSITION. The normalized deviations $n(\vartheta_n^*-\vartheta)$ are bounded in P_ϑ-probability and all the moments of these random variables are bounded uniformly in $\vartheta\in(h,1-h)$.

Proof. Note that for the designs (A1), (A2), (B2) we have

$$|i_0/n - \vartheta| \le 2/n,\quad |\hat{i}/n - \vartheta_n^*| \le 2/n,$$

and thus $|n|\vartheta_n^* - \vartheta| - |\hat{i} - i_0|| \le 4$. Also for these designs the inequalities $nh_0 \le i_0 \le n(1-h_0)+2$ hold almost surely with $h_0=h/2$ if n is large. Hence, Lemma 1.9.2 applies, and for any $C>8$

$$P_\vartheta\left(n|\vartheta_n^* - \vartheta|\ge C\right) \le P_\vartheta\left(|\hat{i}-i_0|\ge C-4\right) \le P_\vartheta\left(|\hat{i}-i_0|\ge C/2\right) \le$$

$$\le \sum_{C/2\le|k|\le n} \pi_k + 4n\exp(-C_\pi(nh_0-3)) \le$$

$$\le \sum_{C/2\le|k|} \pi_k + 4n\exp(-C_\pi(nh_0-3)) \le$$

$$\le 2\exp(-C_\pi(C/2-1)) + 4n\exp(-C_\pi(nh_0-3)) \le \exp(-C_\pi C/4)$$

if n is large enough. This proves the proposition for the case of designs (A1), (A2), and (B2).

For the case of random design (B1) we have

$$P_\vartheta\left(n|\vartheta_n^* - \vartheta|\ge C\right) \le P_\vartheta\left(n|\vartheta_n^* - \vartheta|\ge C,\ N_1>nh_1,\ N_2>nh_1\right) +$$

$$+\ P_\vartheta\left(\ N_1\le nh_1\right) + P_\vartheta\left(\ N_2\le nh_1\right)$$

where $N_1 = \mathrm{card}\{i:\ X_i\in (h/2,\vartheta)\}$ and $N_2 = \mathrm{card}\{i:\ X_i\in (\vartheta,1-h/2)\}$, h_1 is a small positive chosen below. Since $\left\{\ N_1>nh_1,\ N_2>nh_1\ \right\}$ implies $\left\{nh_1\le i_0\le n(1-h_1)+2\right\}$, the first probability is bounded from above by $\exp(-C_\pi C/4)$. To estimate $P_\vartheta\left(\ N_1\le nh_1\right)$ note that the random number N_1 is the sum of i.i.d. binary random variables $\tilde\eta_i$:

$$P(\tilde\eta_i=1)=\vartheta-h/2\ \ge\ h/2 \text{ and } P(\tilde\eta_i=0)=1-(\vartheta-h/2),$$

Applying the Chebyshev exponential inequality, we obtain that for any $z>0$

$$P_\vartheta\left(\ N_1\le nh_1\right) = P_\vartheta\left(-N_1\ge -nh_1\right) \le \exp(znh_1)\ E[\exp(z(-\sum_{i=1}^{n}\tilde\eta_i))] =$$

$$= \exp(znh_1)\ \left(E\ [\exp(-z\tilde\eta_1)]\right)^n = \exp(znh_1)\ \left(1\ -(\vartheta-h/2)(1-e^{-z})\right)^n \le$$

$$\le \exp\left(n(zh_1-(\vartheta-h/2)(1-e^{-z}))\right).$$

Choose, e. g., $z=1$, and let h_1 be such that

$$(\vartheta-h/2)(1-e^{-1})-h_1=c_{14}>0$$

Then we have

$$P_\vartheta\left(\ N_1\le nh_1\right) \le \exp(-c_{14}n).$$

The inequality $P_\vartheta\left(\ N_2\le nh_1\right) \le \exp(-c_{15}n)$, $c_{15}>0$, proves similarly and the proposition follows. □

Proposition 1.9.3 shows that the convergence rate of estimators for the model (1.42) is the same as for the model (1.43) and equals $\psi_n=1/n$. But some good properties of continuous model (1.43) are not preserved for its discrete analogue (1.42), e.g., the limit distribution of the log-likelihood ratio is not described by $Z_0(u)$. One property which is of the particular interest for us concerns the bias $b_n(\vartheta)=E_\vartheta(\vartheta_n^*-\vartheta)$ of the MLE. As it follows from (c), in the continuous-time model (1.43) the properly normalized MLE is asymptotically unbiased:

$$\lim_{n\to\infty} \sup_{h<\vartheta<1-h} E_\vartheta\left(n(\vartheta_n^* - \vartheta)\right) = 0.$$

However, it is clear that this property is not always true under the discrete model (1.42). For example, under the regular grid design (A1) the MLE ϑ_n^* is obviously biased, i.e.

$$\liminf_{n\to\infty} \sup_{h<\vartheta<1-h} E_\vartheta\left(n(\vartheta_n^* - \vartheta)\right) > 0.$$

In fact, ϑ_n^* does not change its mean value when ϑ varies within a bin: $(i-1)/n<\vartheta<i/n$. It appears that the designs (A2), (B1), and (B2) based on randomization are closer to the continuous case: for these designs the MLE is asymptotically unbiased, as follows from the next proposition.

1.9.4. PROPOSITION. The MLE based on observations (1.42) with one of designs (A2), (B1), or (B2) satisfies

$$|E_\vartheta(n(\vartheta_n^* - \vartheta))| \le \exp(-\lambda n) \tag{1.46}$$

with some positive $\lambda>0$ for n large enough; inequality (1.46) holds uniformly in $\vartheta\in(h,1-h)$.

We will apply this property in Chapter 6 only to the design (A2), i.e. to the random shift of the regular design, so the proof of Proposition 1.9.4 is presented only for this case. To prove this proposition we first formulate the following auxiliary result. Define the symmetric probability density function

$$\pi(x) = \pi_k \text{ if } k-1/2<x<k+1/2 \ , \ k=0,\pm1,\ldots .$$

1.9.5. LEMMA. Under the design (A2) there exists a positive constant λ_0 such that for n large

$$\sup_{-\infty<x<+\infty} \left| P_\vartheta\left(n(\vartheta_n^* - \vartheta)\le x\right) - \int_{-\infty}^{x} \pi(y)\,dy \right| \le \exp(-\lambda_0 n)$$

uniformly over $\vartheta\in(h,1-h)$.

Proof. Denote by $\bar\vartheta_n = (X_{i_0-1} + X_{i_0})/2$ the midpoint of the design interval containing ϑ. The two random events are coincide under the design (A2): $\{n(\vartheta_n^*-\bar{\vartheta}_n)=k\} = \{\hat{i}-i_0=k\}$, $1<k+i_0\le n$, while the random variable $n(\bar{\vartheta}_n-\vartheta)$ depends on the design only and is uniformly distributed on $[-1/2,1/2]$. Set $h_0=h/2$, $k_n=[nh_0]$. For any integer k, $|k|\le k_n$, and for any $x\in(-1/2,1/2)$, we have

$$P_\vartheta\Big(n(\vartheta_n^*-\vartheta)\in(k-1/2,k+x)\Big) = P_\vartheta\Big(n(\vartheta_n^*-\bar{\vartheta}_n)=k,\ n(\bar{\vartheta}_n-\vartheta)\in(-1/2,x)\Big) =$$

$$= P_\vartheta\Big(\hat{i}-i_0=k,\ n(\bar{\vartheta}_n-\vartheta)\in(-1/2,x)\Big).$$

Applying Lemma 1.9.2, one has

$$\Big|\ P_\vartheta\Big(n(\vartheta_n^*-\vartheta)\in(k-1/2,k+x)\Big) - \int_{k-1/2}^{k+x} \pi(y)dy\ \Big| \le 2\ \exp(-C_\pi(nh_0-3)).$$

Hence for any x, $x\le k_n+1/2$,

$$\Big|\ P_\vartheta\Big(n(\vartheta_n^*-\vartheta)\le x\Big) - \int_{-\infty}^{x} \pi(y)dy\ \Big| \le$$

$$\le \sum_{k=-k_n}^{[x-0.5]} \Big|P_\vartheta\Big(n(\vartheta_n^*-\vartheta)\in(k-1/2,k+1/2)\Big) - \int_{k-1/2}^{k+1/2}\pi(y)dy\Big| +$$

$$+ \Big|P_\vartheta\Big(n(\vartheta_n^*-\vartheta)\in([x-0.5]+1/2,x)\Big) - \int_{[x-0.5]+1/2}^{x} \pi(y)dy\Big| +$$

$$+ P_\vartheta\Big(n(\vartheta_n^*-\vartheta)\le -k_n-1/2\Big) + \int_{-\infty}^{-k_n-1/2} \pi(y)dy \le$$

$$\le 2(2k_n+1)\ \exp(-C_\pi(nh_0-3)) + P_\vartheta(\hat{i}-i_0<-k_n) + \sum_{k\le -k_n}\pi_k .$$

Due to Lemmas 1.9.1 and 1.9.2 the last two summands decrease exponentially fast as $n\to\infty$. This proves the lemma. □

Proof of Proposition 1.9.4. Applying Lemma 1.9.5, the symmetry of $\pi(x)$, and the obvious property $|n(\vartheta_n^*-\bar{\vartheta}_n)|<n$, we have that uniformly in $\vartheta\in(h,1-h)$ the inequality holds

$$|E_\vartheta(n(\vartheta_n^* - \bar{\vartheta}_n))| \le \int_{-\infty}^{\infty} x\pi(x)\,dx + n\exp(-\lambda_0 n) =$$

$$= n\exp(-\lambda_0 n) \le \exp(-\lambda n)$$

for $0<\lambda\le\lambda_0$ and n large enough. □

From Propositions 1.9.3 and 1.9.4 we get the following.

1.9.6. COROLLARY. Under the assumptions of Proposition 1.9.4 the MLE ϑ_n^* has the next stochastic expansion:

$$\vartheta_n^* = \vartheta + b_n(\vartheta) + n^{-1}\varepsilon_n , \quad n = 1,2,\ldots,$$

where the deterministic bias $b_n(\vartheta)=E_\vartheta(\vartheta_n^*-\vartheta)$ is such that

$$|b_n(\vartheta)| \le \exp(-\lambda n), \ \lambda>0,$$

for n large enough uniformly in $\vartheta\in(h,1-h)$; ε_n's are zero-mean random variables with variances $\sigma_n^2(\vartheta)$ which are bounded uniformly in $\vartheta\in(h,1-h)$ and in $n\ge1$.

Non-Gaussian noise. Now we consider some extensions of the previous results. The first one concerns the non-Gaussian random noise in (1.42).

1.9.7. ASSUMPTION. The random variables ξ_i in (1.42) are conditionally independent for fixed design points $(X_1,\ldots,X_n)$, and

$$E\left(\xi_i|X_1,\ldots,X_n\right) = 0 , \quad E\left(\exp(z\xi_i)|X_1,\ldots,X_n\right) \le \exp(Hz^2),$$

$$i=1,\ldots,n,$$

for any real z with some constant H which does not depend on i, n, and on the design points $(X_1,\ldots,X_n)$.

Note that the random variables ξ_i under Assumption 1.9.7 are not necessarily identically distributed. Also ϑ_n^* is now LSE, not necessarily MLE.

1.9.8. PROPOSITION. If Assumption 1.9.7 holds then Proposition 1.9.3 is true for LSE ϑ_n^*.

Proof. We follow the same lines as in Proposition 1.9.3 and use the notation introduced there. Lemma 1.9.1 was the only part of proof which used explicitly the assumption of Gaussian distribution of ξ_i (respectively, $\bar{\xi}_i$). Assumption 1.9.7 allows to change this part by rather standard technique. In fact, if $\bar{\xi}_i$ satisfy Assumption 1.9.7 then

$$P\left(|\hat{k}| \ge k_0\right) \ge P\left(\sup_{|k|\ge k_0} Z_k > 0 | X_1,\dots,X_n \right) \le$$

$$\le \sum_{k\ge k_0} P\left(\sum_{i=1}^{k} \bar{\xi}_i > k/2 | X_1,\dots,X_n \right) + \sum_{k\ge -k_0} P\left(\sum_{i=-k}^{-1} \bar{\xi}_i > -k/2 | X_1,\dots,X_n \right).$$

Consider for definiteness the first sum and apply Chebyshev's exponential inequality :

$$\sum_{k\ge k_0} P\left(\sum_{i=1}^{k} \bar{\xi}_i > k/2 | X_1,\dots,X_n \right) \le$$

$$\le \sum_{k\ge k_0} \exp(-zk/2)\; E\left[\exp(z \sum_{i=1}^{k} \bar{\xi}_i) | X_1,\dots,X_n \right) \le$$

$$\le \sum_{k\ge k_0} \exp(-zk/2+kHz^2) \le \sum_{k\ge k_0} \exp(-k/(16H)) \le c_{17}\exp(-k_0/(32H))$$

if we choose $z=1/(4H)$ and $c_{17}=(1-\exp(-1/(16H)))^{-1}$. □

<u>Multiplicative model</u>. Proposition 1.9.8 has some direct applications to the models which formally differ from (1.42). One of them is the multiplicative model with observations

$$Y_i = \xi_i \tilde{f}(X_i) , \qquad i=1,\dots,n , \tag{1.47}$$

where the function $\tilde{f}(x)=I(x>\vartheta)-I(x\le\vartheta)$, and ξ_i are i.i.d. binary random variables:

$$\xi_i = \begin{cases} 1 & \text{with probability } \bar{p}_0 \\ -1 & \text{with probability } 1-\bar{p}_0, \end{cases}$$

$$1/2 < \bar{p}_0 < 1.$$

The model (1.47) is related to data transmission through a binary symmetric communication channel (see Chapter 5). Assumption 1.9.7 holds for the multiplicative model (1.47) if we slightly modify the observations:

(1.48) $$Y_i' = \tilde{f}(X_i) + \xi_i' , \qquad i=1,\dots,n ,$$

where $Y_i'=Y_i/E(\xi_i)=Y_i/(2\bar{p}_0-1)$; $\xi_i' = (\xi_i-E(\xi_i))\tilde{f}(X_i)/(2\bar{p}_0-1)$.

Nonparametric model. Assume that we observe

(1.49) $$Y_i = f(X_i) + \xi_i, \quad i = 1,\dots,n,$$

where

$$f(x) = \begin{cases} f_0(x) & \text{if} \quad x\le\vartheta, \\ f_1(x) & \text{if} \quad x>\vartheta, \end{cases}$$

and ξ_i are zero-mean Gaussian random variables as in (1.42) . The values of functions $f_0(x)$ and $f_1(x)$ should be separated from each other in a certain way. For example, it is sufficient to assume that

(1.50) $$0 \le f_0(x) \le a < b \le f_1(x) \le 1$$

for some constants a and b. The model (1.49) can be reduced to the multiplicative form (1.47) if, instead of (1.49), we consider the new observations Y_i' defined by

(1.51) $$Y_i' = \begin{cases} 1 & \text{if} \quad Y_i > (a+b)/2, \\ -1 & \text{otherwise} . \end{cases}$$

It is easy to see that $Y_i' = \xi_i' \tilde{f}(X_i)$ where ξ_i' are binary variables, and $\tilde{f}$ is the function from (1.47).

$$\xi_i' = \begin{cases} 1 & \text{with probability } \bar{p}_i, \\ -1 & \text{with probability } 1-\bar{p}_i, \end{cases}$$

where

$$p_i = P\Big(\xi_i+f(X_i) > (a+b)/2\Big) \ge P\Big(\xi_i > (b-a)/2\Big) > 1/2$$

if $X_i \ge \vartheta$. Since the nonparametric function $f(x)$ in (1.49) is unknown, this model cannot be reduced to the additive form (1.48). Nevertheless, the LSE obtained from observations (1.51) satisfies the statement of Proposition 1.9.3. What is more, the good properties of LSE hold uniformly over the nonparametric set of functions defined by (1.50).

For the estimation of change-point parameter ϑ in the continuous-time analogue of (1.49) see Korostelev (1987a,b).

CHAPTER 2. MINIMAX LOWER BOUNDS

2.1. GENERAL STATISTICAL MODEL AND MINIMAX RATES OF CONVERGENCE

In Chapter 1 we found the convergence rates of some estimators in nonparametric regression and in the change-point problem. The purpose of this chapter is to show that these rates of convergence cannot be improved by any other estimators.

We would like to study the bounds on the accuracy of estimators in these two statistical problems in parallel, though it may seem they have few common features. To realize this plan we embed them into a more general framework. Consider these particular models as examples of the general statistical model.

2.1.1. GENERAL STATISTICAL MODEL. The general statistical model includes the following elements.

(i) The number of observations n. In this book we use asymptotic approach, i.e. we obtain all the results under the assumption that n tends to infinity.

(ii) The subject of estimation, or unknown "parameter", which will be denoted by ϑ . The meaning of ϑ depends on the particular model: in nonparametric regression $\vartheta=f(x)$ is a regression function, while in the change-point problem ϑ is a time-point, i.e. the one-dimensional parameter. In next chapters related to image estimation we consider statistical models where ϑ is a closed set on the plane.

In general ϑ is assumed to be an element of a space Θ endowed with some pseudometric $d(\cdot,\cdot)$.(see Section 1.7 for the examples of pseudometrics).

As a rule, the statistician has some additional information on the parameter ϑ. In other words, the statistician knows that ϑ belongs to some subset Σ of Θ. For example, in regression problem Σ can be the smoothness class of functions $\Sigma(\beta,L)$ defined in Section 1.2.

(iii) The vector of observations $X^{(n)}$ which is supposed to be an element of a measurable space $(\mathfrak{X}^{(n)}, \mathfrak{F}^{(n)})$. In general, this can be an arbitrary measurable space. However, in our examples $X^{(n)}$ is a vector in some Euclidean space.

Thus, for nonparametric regression model with regular design (see Section 1.2) we have

$$Y_i = f(i/n) + \xi_i \ , \ i = 1, \ldots, n.$$

Here the vector of observations is

$$X^{(n)} = (Y_1, \ldots, Y_n) \in R^n,$$

and $\mathfrak{X}^{(n)} = R^n$ .If the design is random and $Y_i = f(X_i) + \xi_i$, where X_i are i.i.d. in the interval [0,1], i = 1,..., n, then

$$X^{(n)} = ((X_1, Y_1), \ldots, (X_n, Y_n)) \in ([0,1]xR^1)^n,$$

and $\mathfrak{X}^{(n)} = ([0,1]xR^1)^n$.

(iv) The <u>family of distributions</u> $P_\vartheta^{(n)} = P_\vartheta^{(n)}(X^{(n)})$ generated by observations $X^{(n)}$ in $(\mathfrak{X}^{(n)}, \mathfrak{F}^{(n)})$. We shall often write for brevity $P_\vartheta = P_\vartheta^{(n)}$. It is not a strong restriction to assume that there is a measure $\nu = \nu^{(n)}$ on the space $(\mathfrak{X}^{(n)}, \mathfrak{F}^{(n)})$ which dominates all the measures P_ϑ so that the density

$$p(x,\vartheta) = p^{(n)}(x,\vartheta) = \left(dP_\vartheta/d\nu\right)(x)$$

is correctly defined for $x \in \mathfrak{X}^{(n)}$ and $\vartheta \in \Sigma \subset \Theta$ (see Ibragimov and Khasminskii (1981)).

(v) <u>Estimators</u> $\hat{\vartheta}_n = \hat{\vartheta}_n(X^{(n)})$ which are measurable mappings with respect to observations, i. e. $\hat{\vartheta}_n : \mathfrak{X}^{(n)} \to \Theta$. Remark: we don't assume that the estimators $\hat{\vartheta}_n$ belong to the set Σ. For example, an estimator of a smooth regression function is not necessarily a smooth function.

(vi) The <u>risk function</u>

$$R(\hat{\vartheta}_n, \vartheta) = R(\hat{\vartheta}_n, \vartheta, \psi_n) = E_\vartheta\left(w(\psi_n^{-1} d(\hat{\vartheta}_n, \vartheta)) \right).$$

(cf. Section 1.7). Here w(u) is a loss function, ψ_n is a positive sequence characterizing the rate of convergence, E_ϑ is the expectation with respect to P_ϑ. We deal only with the simplest loss functions w discussed in example 1.7.1. The first one is the power loss function $w(u)=|u|^a$ with a positive a. In most examples a=2 (the squared loss). The second one is the indicator loss function $w(u)=I(|u|\geq C)$. In this case

$$R(\hat{\vartheta}_n,\vartheta,\psi_n) = P_\vartheta\left(\psi_n^{-1}d(\hat{\vartheta}_n,\vartheta) \geq C\right).$$

2.1.2. MINIMAX RATES OF CONVERGENCE AND OPTIMAL ESTIMATORS. In Section 1.7 we argued for the minimax approach as one of the proper ways of comparing between nonparametric regression estimators. This approach can be extended to the general statistical model. We start with some basic definitions which are due to Ibragimov and Khasminskii (1980a,b, 1981, 1982), Bretagnolle and Huber (1979), Stone (1980, 1982) and Birgé (1983).

Define the <u>maximal risk</u> of an estimator $\hat{\vartheta}_n$ on the set Σ as

$$r(\hat{\vartheta}_n) = r(\hat{\vartheta}_n,\psi_n) = \sup_{\vartheta\in\Sigma} R(\hat{\vartheta}_n,\vartheta,\psi_n).$$

2.1.3. DEFINITION. The positive sequence ψ_n is called <u>lower rate of convergence</u> for the set Σ in pseudometric $d(\cdot,\cdot)$ if

$$\liminf_{n\to\infty} \inf_{\hat{\vartheta}_n} r(\hat{\vartheta}_n,\psi_n) \geq C_0 \tag{2.1}$$

with some positive C_0. Here and later $\inf_{\hat{\vartheta}_n}$ denotes the infimum over all estimators.

Inequality (2.1) is a kind of negative statement that says that the estimators of parameter ϑ cannot converge to ϑ very fast (in the sense of convergence of their maximal risks).

Note that the lower rate of convergence is not uniquely defined. Thus, if ψ_n satisfies (2.1) then, at least for our case of power or indicator loss functions, any positive sequence $\{\psi_n'\}$, $\psi_n' \leq \psi_n$, also satisfies (2.1).

2.1.4. DEFINITION. The positive sequence ψ_n is called <u>minimax rate of convergence</u> for the set Σ in pseudometric $d(\cdot,\cdot)$ if it is lower rate of convergence, and there exists an estimator ϑ_n^* such that

$$\limsup_{n\to\infty} r(\vartheta_n^*,\psi_n) \leq C_1 \tag{2.2}$$

for some constant C_1. If so, ϑ_n^* is called <u>optimal</u> <u>estimator</u>.

The motivation for Definitions 2.1.3 and 2.1.4 is clear from the following example.

2.1.5. EXAMPLE. Consider the case of squared loss function. Then (2.1) means that for n large and <u>for any estimator</u> $\hat{\vartheta}_n$

$$\sup_{\vartheta\in\Sigma} E_\vartheta(d^2(\hat{\vartheta}_n,\vartheta)) \geq C_0'\psi_n^2 \tag{2.3}$$

while (2.2) has the form

$$\sup_{\vartheta\in\Sigma} E_\vartheta(d^2(\vartheta_n^*,\vartheta)) \leq C_1'\psi_n^2 \tag{2.4}$$

where $0<C_0'<C_0$, $C_1<C_1'<\infty$ (for n large). If both (2.3) and (2.4) are satisfied then the estimator ϑ_n^* has the fastest convergence rate among all possible estimators (in the sense of convergence of maximal risks). This fastest convergence rate ψ_n is the minimax rate according to Definition 2.1.4. In the following we use the words <u>S-lower rate of convergence</u>, <u>S-minimax rate of convergence</u>, and <u>S-optimal estimators</u> to denote respectively lower rates of convergence, minimax rates of convergence and optimal estimators in case of <u>squared</u> loss function. The same objects for <u>absolute value</u> loss function are called <u>A-lower rate of convergence</u>, <u>A-minimax rate of convergence</u>, and <u>A-optimal estimators</u>.

The minimax rate of convergence is not unique: it is defined to within a constant. Indeed, if ψ_n is a minimax rate of convergence and we consider the indicator or power loss functions then for any sequence ψ_n' which satisfies

$$0 < \liminf_{n\to\infty} \left(\psi_n / \psi_n'\right) \leq \limsup_{n\to\infty} \left(\psi_n / \psi_n'\right) < \infty$$

conditions (2.1) and (2.2) are also fulfilled. We call such sequences ψ_n, ψ_n' equivalent in order.

2.1.6. DEFINITION. Let ψ_n be the minimax rate of convergence for the set Σ in pseudometric $d(\cdot,\cdot)$. Then

$$r_n^* = \inf_{\hat{\vartheta}_n} r(\hat{\vartheta}_n,\psi_n) = \inf_{\hat{\vartheta}_n} \sup_{\vartheta\in\Sigma} R(\hat{\vartheta}_n,\vartheta,\psi_n).$$

is the <u>minimax risk</u> for the set Σ in pseudometric $d(\cdot,\cdot)$.

It is clear that (2.1) gives the asymptotic lower bound for the minimax risk. Hence inequality (2.1) is usually called <u>minimax lower bound</u>. On the other hand, (2.2) gives the upper bound for the minimax risk since for any estimator ϑ_n^*

$$r_n^* \le r(\vartheta_n^*, \psi_n).$$

Looking again at Example 2.1.5 we see why the sequence ψ_n satisfying (2.3) and (2.4) has got the name "minimax rate of convergence". Note that for this example r_n^* and ψ_n^2 are equivalent in order. In other words, ψ_n characterizes the convergence rate of the minimax risk. The same is true for other examples of power loss functions.

However, the Definition 2.1.4 is not well suited to the case of bounded loss functions such as the indicator loss. In fact, if $w(u)=I(|u|\ge C)$ then (2.2) is trivially satisfied for all estimators (put $C_1=1$).

Let us modify the definitions to make them acceptable for the case of indicator loss. In this we follow Stone (1980) who proposed to consider the asymptotics both in n and in C. Introduce the following notation where the dependence of the risk function on C is explicit:

$$r(\hat{\vartheta}_n) = r(\hat{\vartheta}_n, \psi_n, C) = \sup_{\vartheta\in\Sigma} P_\vartheta\left(\psi_n^{-1} d(\hat{\vartheta}_n, \vartheta) \ge C \right).$$

2.1.7. DEFINITION. The positive sequence ψ_n is called <u>I-lower rate of convergence</u> for the set Σ in pseudometric $d(\cdot,\cdot)$ if there exists $C>0$ such that

$$\liminf_{n\to\infty} \inf_{\hat{\vartheta}_n} r(\hat{\vartheta}_n, \psi_n, C) \ge p_0 \tag{2.5}$$

with some positive p_0.

In other words, the I-lower rate ψ_n is such that for any estimator $\hat{\vartheta}_n$ the sequence $\psi_n^{-1} d(\hat{\vartheta}_n, \vartheta)$ is bounded away from 0 in P_ϑ-probability at least for some value of $\vartheta\in\Sigma$ as n tends to ∞. In fact, Definition 2.1.7 is a weaker version of Definition 2.1.3, with the additional choice of suitable $C>0$. As (2.1), inequality (2.5) is called <u>minimax lower bound</u>.

2.1.8. DEFINITION. The positive sequence ψ_n is called <u>I-minimax rate of convergence</u> for the set Σ in pseudometric $d(\cdot,\cdot)$ if it is I-lower rate of convergence, and there exists an estimator ϑ_n^* such that

$$(2.6) \qquad \lim_{C\to\infty} \limsup_{n\to\infty} r(\vartheta_n^*, \psi_n, C) = 0 .$$

If so, ϑ_n^* is called <u>I-optimal estimator</u>.

The relation (2.6) means that the normalized deviation $\psi_n^{-1} d(\vartheta_n^*, \vartheta)$ is bounded in P_ϑ-probability uniformly over $\vartheta\in\Sigma$ as n tends to ∞.

For the majority of statistical models I-minimax rates and I-optimal estimators are the same as S-minimax rates and S-optimal estimators respectively. As we see later, this is true for the models considered in this book.

In this chapter we describe general techniques of proving minimax lower bounds (2.1) and (2.5), and apply them to the problems of nonparametric regression and change-point estimation. As a consequence, we show that the estimators studied in Chapter 1 are optimal.

2.2. THE BASIC IDEA

Inequality (2.5) can be rewritten in the equivalent expanded form: for any estimator $\hat{\vartheta}_n(X^{(n)})$ and for n sufficiently large

$$(2.7) \qquad \sup_{\vartheta\in\Sigma} \int_{\mathfrak{X}^{(n)}} I(d(\hat{\vartheta}_n(X^{(n)}), \vartheta) \ge C\psi_n)\; P_\vartheta(dX^{(n)}) \ge p_0 .$$

The main difficulty in analyzing the extremal problem in the left-hand side of (2.7) is that the set of all estimators $\hat{\vartheta}_n(X^{(n)})$ is too large. Explicit solution of this extremal problem is rarely feasible. Nevertheless, there exists a simple idea which allows to get rid of an arbitrary measurable function $\hat{\vartheta}_n(X^{(n)})$ in (2.7). First, note that the supremum $\sup_{\vartheta\in\Sigma}$ in (2.7) is bounded from below by the maximum over some finite subset of Σ. Consider the simplest case of a subset containing only two elements, ϑ' and ϑ''. Then

$$\text{(2.8)} \qquad \sup_{\vartheta\in\Sigma} P_\vartheta\Big(\psi_n^{-1} d(\hat\vartheta_n,\vartheta) \ge C \Big) \ge \max_{\vartheta=\vartheta',\vartheta''} P_\vartheta\Big(\psi_n^{-1} d(\hat\vartheta_n,\vartheta) \ge C \Big),$$

and the problem is to bound the right-hand side of (2.8) from below. Now we explain how to do this.

To start with, consider the degenerate case. Assume that $P_{\vartheta'} \equiv P_{\vartheta''}$ for some $\vartheta'=\vartheta_n'$ and $\vartheta''=\vartheta_n''$ such that $d(\vartheta',\vartheta'') = 2s_n > 0$. It is intuitively clear that observations $X^{(n)}$ give no additional information on the "true" value of ϑ. Hence all we are able to do is merely to guess the value of ϑ, with probability of error at least 1/2. To show this in a formal way note that for any estimator $\hat\vartheta_n$ the triangle inequality for pseudometric $d(\cdot,\cdot)$ guarantees the following implication :

$$\text{(2.9)} \qquad \Big\{ d(\hat\vartheta_n,\vartheta'') < s_n \Big\} \subseteq \Big\{ d(\hat\vartheta_n,\vartheta') \ge s_n \Big\} .$$

Hence,

$$\text{(2.10)} \quad 1 = P_{\vartheta''}\Big(d(\hat\vartheta_n,\vartheta'') < s_n\Big) + P_{\vartheta''}\Big(d(\hat\vartheta_n,\vartheta'') \ge s_n\Big) =$$

$$= P_{\vartheta'}\Big(d(\hat\vartheta_n,\vartheta'') < s_n\Big) + P_{\vartheta''}\Big(d(\hat\vartheta_n,\vartheta'') \ge s_n\Big) \le$$

$$\le P_{\vartheta'}\Big(d(\hat\vartheta_n,\vartheta') \ge s_n\Big) + P_{\vartheta''}\Big(d(\hat\vartheta_n,\vartheta'') \ge s_n\Big).$$

Consequently, the last probabilities cannot simultaneously be less than 1/2 , i. e. the minimax lower bound holds

$$\text{(2.11)} \qquad \max_{\vartheta=\vartheta',\vartheta''} P_\vartheta \Big(d(\hat\vartheta_n,\vartheta) \ge s_n \Big) \ge 1/2 .$$

What should be assumed to extend this idea to the non-degenerate case? Denote the likelihood ratio by

$$\Lambda(\vartheta',\vartheta'') = \Lambda(\vartheta',\vartheta'';X^{(n)}) = p(X^{(n)},\vartheta')/p(X^{(n)},\vartheta'')$$

where $p(x,\vartheta)$ is the density introduced in Section 2.1.1 (iv). The fact that $P_{\vartheta'} \equiv P_{\vartheta''}$ may be expressed in terms of the likelihood ratio as

$$(2.12) \qquad \Lambda(\vartheta',\vartheta'';X^{(n)}) = 1 \quad \text{almost surely with respect to } P_{\vartheta''}.$$

Indeed, if (2.12) is true then for any random event A we have

$$P_{\vartheta'}(A) = E_{\vartheta''}\Big[\Lambda(\vartheta',\vartheta'';X^{(n)}) \, I(X^{(n)} \in A) \Big] = P_{\vartheta''}(A) .$$

The following assumption generalizes (2.12) .

2.2.1. ASSUMPTION . Let $\vartheta' = \vartheta'_n$, $\vartheta'' = \vartheta''_n$, and let

$$(2.13) \qquad P_{\vartheta''}\Big(\Lambda(\vartheta',\vartheta'';X^{(n)}) > e^{-\lambda} \Big) \ge p$$

for some $\lambda > 0$ and $p > 0$ which are independent of n .

2.2.2. PROPOSITION . If Assumption 2.2.1 is satisfied and $d(\vartheta',\vartheta'') = 2s_n > 0$ then for any estimator $\hat{\vartheta}_n$ the minimax lower bound holds

$$(2.14) \qquad \max_{\vartheta=\vartheta',\vartheta''} P_{\vartheta}\Big(d(\hat{\vartheta}_n,\vartheta) \ge s_n \Big) \ge p\, e^{-\lambda}/2.$$

Proof. Similarly to (2.10)

$$P_{\vartheta'}\Big(d(\hat{\vartheta}_n,\vartheta') \ge s_n\Big) + P_{\vartheta''}\Big(d(\hat{\vartheta}_n,\vartheta'') \ge s_n\Big) =$$

$$= E_{\vartheta''}\Big[\Lambda(\vartheta',\vartheta'';X^{(n)}) I(d(\hat{\vartheta}_n,\vartheta') \ge s_n)\Big] + P_{\vartheta''}\Big(d(\hat{\vartheta}_n,\vartheta'') \ge s_n\Big) \ge$$

$$\ge E_{\vartheta''}\Big[\Lambda(\vartheta',\vartheta'';X^{(n)}) \, I(\Lambda(\vartheta',\vartheta'';X^{(n)}) > e^{-\lambda}) \, I(d(\hat{\vartheta}_n,\vartheta') \ge s_n)\Big] +$$

$$+ P_{\vartheta''}\Big(d(\hat{\vartheta}_n,\vartheta'') \ge s_n \Big) \ge$$

$$\ge e^{-\lambda} P_{\vartheta''}\Big((\Lambda(\vartheta',\vartheta'';X^{(n)}) > e^{-\lambda}) \cap (d(\hat{\vartheta}_n,\vartheta') \ge s_n) \Big) +$$

$$+ P_{\vartheta''}\left(d(\hat{\vartheta}_n, \vartheta'') \ge s_n \right).$$

Using (2.9) once more, we finally obtain

$$P_{\vartheta'}\left(d(\hat{\vartheta}_n, \vartheta') \ge s_n\right) + P_{\vartheta''}\left(d(\hat{\vartheta}_n, \vartheta'') \ge s_n\right) \ge$$

$$\ge e^{-\lambda} P_{\vartheta''}\left((\Lambda(\vartheta', \vartheta''; X^{(n)}) > e^{-\lambda}) \cap (d(\hat{\vartheta}_n, \vartheta'') < s_n) \right) +$$

$$+ P_{\vartheta''}\left(d(\hat{\vartheta}_n, \vartheta'') \ge s_n \right) \ge e^{-\lambda} P_{\vartheta''}\left(\Lambda(\vartheta', \vartheta''; X^{(n)}) > e^{-\lambda}\right) \ge$$

$$\ge p\, e^{-\lambda} = p_0 > 0 ,$$

which proves (2.14) similarly to (2.11) .□

Thus, we have reduced the problem of proving minimax lower bounds to the problem of choosing two elements ϑ', ϑ'' of the set Σ satisfying certain conditions. For particular statistical models the choice of ϑ', ϑ'' is a matter of art. Of course, one is interested to find the least favorable pair $(\vartheta', \vartheta'')$, i.e. the pair for which the distance s_n is the largest (at least in some asymptotic sense) among all pairs satisfying Assumption 2.2.1. There exist some recipes of choosing such pairs for a number of examples. We discuss them in Section 2.4.

2.3. DISTANCES BETWEEN DISTRIBUTIONS

Inequality (2.13) in Assumption 2.2.1 may be interpreted as the fact that the distributions $P_{\vartheta'}$ and $P_{\vartheta''}$ are close to each other. For example, if $P_{\vartheta'} = P_{\vartheta''}$ then (2.13) holds for any $\lambda>0$. The same is true if the likelihood ratio is close to 1 in some probabilistic sense. However, (2.13) does not propose an easy way of comparing between different distributions on the common scale. To make such comparison possible the closeness measures of distance type are used. In this section we discuss briefly some of them.

Total variation distance. Let P and Q be probability measures on $(\mathfrak{X}^{(n)}, \mathcal{F}^{(n)})$ having densities p and q with respect to $\nu = \nu^{(n)}$. The total variation distance V(P,Q) between P and Q is defined as

$$V(P,Q) = \sup_{A \in \mathcal{F}^{(n)}} | P(A) - Q(A) |.$$

2.3.1. PROPOSITION. The total variation distance has the following representation :

$$(2.15) \qquad V(P,Q) = 1 - \int_{\mathfrak{X}^{(n)}} \min\Big[p(x), q(x)\Big] \nu(dx) =$$

$$= \frac{1}{2} \int_{\mathfrak{X}^{(n)}} |p(x) - q(x)| \nu(dx).$$

Proof. For the sake of brevity we omit arguments of functions. First of all, note that

$$V(P,Q) = \sup_A |P(A) - Q(A)| = \sup_A \left| \int_A p \, d\nu - \int_A q \, d\nu \right| \geq$$

$$\geq \int_{p \geq q} p \, d\nu - \int_{p \geq q} q \, d\nu.$$

The last inequality holds if we take the particular set $A = A_0 = \{x: p(x) \geq q(x)\}$. It leads to

$$(2.16) \qquad V(P,Q) \geq 1 - \int_{p<q} p \, d\nu - \int_{p \geq q} q \, d\nu = 1 - \int \min\Big[p,q \Big] d\nu.$$

On the other hand, let A be an arbitrary set. Without loss of generality we may suppose that $P(A) \geq Q(A)$. Then

$$(2.17) \qquad 0 \leq P(A) - Q(A) = \int_{A \cap A_0} p \, d\nu + \int_{A \cap \bar{A}_0} p \, d\nu -$$

$$- \left(\int_{A \cap A_0} q \, d\nu + \int_{A \cap \bar{A}_0} q \, d\nu \right) \le \int_{A \cap A_0} (p - q) \, d\nu \le$$

$$\le \int_{A_0} (p - q) \, d\nu = 1 - \int \min\left[p, q \right] d\nu .$$

Inequalities (2.16) and (2.17) prove (2.15). □

2.3.2. ASSUMPTION . Let $\vartheta' = \vartheta'_n$, $\vartheta'' = \vartheta''_n$, and let for some positive V_0 independent of n the following inequality hold :

$$V\left(P_{\vartheta'}(X^{(n)}),\ P_{\vartheta''}(X^{(n)}) \right) \le V_0 < 1.$$

This assumption is sufficient to get the following minimax lower bound.

2.3.3. PROPOSITION . If Assumption 2.3.2 is fulfilled and $d(\vartheta', \vartheta'') = 2s_n > 0$, then for any estimator $\hat{\vartheta}_n$ the minimax lower bound holds

$$\max_{\vartheta = \vartheta', \vartheta''} P_\vartheta \left(d(\hat{\vartheta}_n, \vartheta) \ge s_n \right) \ge (1 - V_0)/2. \tag{2.18}$$

<u>Proof</u>. Let $p(x, \vartheta')$ and $p(x, \vartheta'')$ be the densities of $P_{\vartheta'}$ and $P_{\vartheta''}$ with respect to $\nu = \nu^{(n)}$. For any estimator $\hat{\vartheta}_n$ we have from (2.9):

$$P_{\vartheta'}\left(d(\hat{\vartheta}_n, \vartheta') \ge s_n\right) + P_{\vartheta''}\left(d(\hat{\vartheta}_n, \vartheta'') \ge s_n\right) \ge \tag{2.19}$$

$$\ge P_{\vartheta'}\left(d(\hat{\vartheta}_n, \vartheta'') < s_n\right) + P_{\vartheta''}\left(d(\hat{\vartheta}_n, \vartheta'') \ge s_n\right) \ge$$

$$\ge \int_{d(\hat{\vartheta}_n, \vartheta'') < s_n} \min[p(x, \vartheta'), p(x, \vartheta'')] d\nu +$$

$$+ \int_{d(\hat{\vartheta}_n,\vartheta'')\ge s_n} \min[p(x,\vartheta'),p(x,\vartheta'')]d\nu =$$

$$= \int \min\,[p(x,\vartheta'),p(x,\vartheta'')]\;d\nu \;=\; 1 - V\left(P_{\vartheta'},P_{\vartheta''}\right) \ge 1 - V_0 > 0.$$

The last equality follows from the property (2.15). The rest of proof is as in Proposition 2.2.2. □

Hellinger distance. Another useful distance between probability measures is the Hellinger distance $H(P,Q)$ defined by

(2.20) $$H^2(P,Q) = \int_{\mathfrak{X}^{(n)}} \left(\sqrt{p(x)} - \sqrt{q(x)} \right)^2 \nu(dx).$$

For each pair of measures P and Q the Hellinger distance satisfies

$$0 \le H^2(P,Q) \le 2 \;.$$

2.3.4. REMARK. If Q is absolutely continuous with respect to P, i.e. the density dQ/dP exists, then

(2.21) $$H^2(P,Q) = \int_{\mathfrak{X}^{(n)}} \left((dQ/dP(x))^{1/2} - 1 \right)^2 P(dx) \;.$$

2.3.5. PROPOSITION. The Hellinger distance and the total variation distance are related to each other in the following way:

(2.22) $$\tfrac{1}{2}H^2(P,Q) \le V(P,Q) \le \left(1- [1 - \tfrac{1}{2}H^2(P,Q))]^2 \right)^{1/2}.$$

Proof. Denote $H = H(P,Q)$, $V = V(P,Q)$, and

$$r = \int (pq)^{1/2} d\nu \;.$$

Since

$$r \ge \int \min[p,q]\;d\nu$$

we have

$$H^2/2 = 1 - r \le 1 - \int \min[p,q]\ d\nu = V.$$

The right-hand inequality in (2.22) is a consequence of (2.15) and of the Cauchy-Schwartz inequality :

$$V = 1 - \int \min\Big[\ p,q\ \Big]\ \nu(dx) = \frac{1}{2}\int |p-q|\nu(dx) =$$

$$= \frac{1}{2}\int |\sqrt{p} - \sqrt{q}\,|(\sqrt{p} + \sqrt{q}\,)\ \nu(dx) \le$$

$$\le \frac{1}{2}\left(\int(\sqrt{p} - \sqrt{q}\,)^2\nu(dx)\right)^{1/2}\left(\int(\sqrt{p} + \sqrt{q}\,)^2\nu(dx)\right)^{1/2} =$$

$$= \frac{1}{2}\Big(H^2(P,Q)(4 - H^2(P,Q))\Big)^{1/2} = \Big(1 - (1-\frac{1}{2}H^2(P,Q))^2\Big)^{1/2}$$

which proves (2.22). □

2.3.6. COROLLARY (Le Cam's inequality).

$$\int \min[p,q]\ d\nu \ge \frac{1}{2}\left(\int (pq)^{1/2}d\nu\right)^2 .$$

<u>Proof</u>. By (2.15) the right-hand inequality in (2.22) can be written as

$$1 - \int \min[p,q]d\nu \le (1-r^2)^{1/2}$$

which is equivalent to

$$2\int \min[p,q]d\nu - \left(\int \min[p,q]d\nu\right)^2 \ge r^2 = \left(\int (pq)^{1/2}d\nu\right)^2 .$$

Hence, the corollary follows. □

The following assumption and proposition are in analogy to Assumption 2.3.2 and Proposition 2.3.3 in terms of the Hellinger distance.

2.3.7. ASSUMPTION. Let $\vartheta' = \vartheta'_n$, $\vartheta'' = \vartheta''_n$, and let for some positive H_0 independent of n the following inequality hold :

$$(2.23) \qquad H^2\Big(P_{\vartheta'}, P_{\vartheta''} \Big) \le H_0^2 < 2.$$

2.3.8. PROPOSITION. If Assumption 2.3.7 is fulfilled and $d(\vartheta',\vartheta'')=2s_n>0$, then for any estimator $\hat{\vartheta}_n$ the minimax lower bound holds

$$(2.24) \qquad \max_{\vartheta=\vartheta',\vartheta''} P_\vartheta\Big(d(\hat{\vartheta}_n,\vartheta) \ge s_n\Big) \ge \frac{1}{4}\,(1-H_0^2/2)^2 > 0.$$

Proof. From (2.19) we find that the left-hand side of (2.24) is greater or equal to $\frac{1}{2}\int \min[p(x,\vartheta'),p(x,\vartheta'')]d\nu$. Hence, Corollary 2.3.6 together with (2.23) give (2.24). □

2.4. EXAMPLES

Here we give some applications of the minimax lower bounds of Section 2.2.

2.4.1. EXAMPLE. Nonparametric regression, regular design.

Let the vector of observations be $X^{(n)}=(Y_1,\dots,Y_n)$ where

$$(2.25) \qquad Y_i = f(i/n) + \xi_i , \quad i = 1,\dots, n ,$$

where ξ_i are independent zero-mean Gaussian random variables with variance $\sigma^2=1$, and the regression function $f(x)$ belongs to the class $\Sigma(\beta,L)$ introduced in Section 1.2. The subject of estimation is the regression function $f(x)$, so we denote $\vartheta=f(\cdot)$. If we want to estimate the regression function at a given point x_0, $0<x_0<1$, then the appropriate pseudometric is $d(f,g)=|f(x_0)-g(x_0)|$.

Let us prove that for this statistical model the I-minimax lower bound is true with $\psi_n=n^{-\beta/(2\beta+1)}$.

We apply the techniques of Section 2.2. This means that we find two elements $\vartheta',\vartheta''\in\Sigma(\beta,L)$ such that the corresponding likelihood ratio satisfies our assumptions, and then apply (2.8) and Proposition 2.2.2. Define $\vartheta''=\vartheta''_n$ as the function which equals to 0 everywhere in [0,1]:

$$\vartheta'' = \vartheta''_n = 0,$$

and choose $\vartheta' = \vartheta'_n = f_{1n}(\cdot)$ so that

(i) For each n the function $f_{1n}(x)$ belongs to $\Sigma(\beta, L)$;

(ii) The likelihood ratio $\Lambda(\vartheta'_n, \vartheta''_n; X^{(n)})$ satisfies Assumption 2.2.1.

Note that $2s_n = |f_{1n}(x_0)|$. To get the better lower bound one should choose the value $|f_{1n}(x_0)|$ as large as possible under the restrictions (i) and (ii).

Let us start with (ii) . The likelihood ratio for Gaussian observations (2.25) has the explicit expression :

(2.26) $$\Lambda(\vartheta'_n, \vartheta''_n; X^{(n)}) =$$

$$= \exp\left\{ -\frac{1}{2} \sum_{i=1}^{n} (Y_i - f_{1n}(i/n))^2 + \frac{1}{2} \sum_{i=1}^{n} Y_i^{\,2} \right\} =$$

$$= \exp\left\{ \sum_{i=1}^{n} f_{1n}(i/n)\, Y_i - \frac{1}{2} \sum_{i=1}^{n} f_{1n}^2(i/n) \right\}.$$

Hence, the logarithm of this likelihood ratio remains bounded from below as n tends to infinity if and only if the sum $\sum_{i=1}^{n} f_{1n}^2(i/n)$ is bounded from above. But this sum is approximately equal to the normalized integral, i.e.

$$\sum_{i=1}^{n} f_{1n}^2(i/n) = n\,(1 + o(1)) \int_0^1 f_{1n}^2(x)\,dx ,$$

as $n \to \infty$. This implies that the integral $\int_0^1 f_{1n}^2(x)\,dx$ should be of the order $O(n^{-1})$ as $n \to \infty$. To guarantee the properties (i) and (ii) it is mathematically most convenient to take the smooth but spiky functions $f_{1n}(x)$ satisfying the following conditions: they are identically zero outside of the interval (x_0-h_n, x_0+h_n) where h_n is a sequence of positive numbers vanishing as n tends to infinity, and the maximal value of $|f_{1n}(x_0)|$ for such a function has the order $O(h_n^{\beta})$ (the last is necessary to ensure that f_{1n} belongs to the class $\Sigma(\beta,L)$). The parameter h_n here should not be confused with the bandwidth of locally-polynomial estimators (cf. Sections 1.3, 1.4). We used this notation to emphasize the fact that the functions f_{1n} may be regarded as rescaled kernels. The relevant choice of h_n here will be also the same as the optimal bandwidth choice discussed in Section 1.6. To construct $f_{1n}(x)$ we implement a "basic" function $\phi(x)$ which is infinitely many times continuously differentiable and compactly supported.

2.4.2. DEFINITION. The function $\phi:R^1 \to R^1$ is called basic function for the class $\Sigma(\beta,L)$ if

(a) ϕ is infinitely many times continuously differentiable on R^1,

(b) $\phi(x)=0$ if $x\notin(-\frac{1}{2},\frac{1}{2})$,

(c) $\phi(x)>0$ if $x\in(-\frac{1}{2},\frac{1}{2})$,

(d) $\max_x |\phi^{(k+1)}(x)| \le L$ where $k=\lfloor\beta\rfloor$.

It is easy to see that there exists a function $\phi=\phi_0$ satisfying 2.4.2 (a)-(c) such that $\bar{\phi}_0 = \max_x |\phi_0^{(k+1)}(x)| \ne 0$. Then $\phi(x)=L\phi_0(x)/\bar{\phi}_0$ is a basic function.

Define $f_{1n}(x)$ as

$$f_{1n}(x) = h_n^{\beta}\phi(h_n^{-1}(x - x_0)). \tag{2.27}$$

2.4.3. PROPOSITION. If ϕ is a basic function for $\Sigma(\beta,L)$ then $f_{1n}\in\Sigma(\beta,L)$.

Proof. It suffices to show

$$|f_{1n}^{(k)}(x) - f_{1n}^{(k)}(x')| \le L|x - x'|^{\beta-k}, \quad k=\lfloor\beta\rfloor, \tag{2.28}$$

only for x,x' such that $|x - x'| \le h_n$ since the support of f_{1n} is the interval of length h_n. We have

$$f_{1n}^{(k)}(x) = h_n^{\beta-k}\,\phi^{(k)}(h_n^{-1}(x - x_0)).$$

Using 2.4.2 (d) we find

$$|\phi^{(k)}(z) - \phi^{(k)}(z')| \le L|z - z'|, \quad \forall\ z,z' \in R^1.$$

Since $\beta - k \le 1$ then

$$|\phi^{(k)}(z) - \phi^{(k)}(z')| \le L|z - z'|^{\beta-k}$$

if $|z-z'|\le 1$. Denoting $z = h_n^{-1}(x-x_0)$, $z' = h_n^{-1}(x'-x_0)$ we obtain (2.28).

□

Using the Riemann summation theorem is easy to verify that if $nh_n \to \infty$ then

$$\sum_{i=1}^{n} f_{1n}^2(i/n) = nh_n^{2\beta+1}(\sigma_0^2 + o(1))$$

as $n \to \infty$ where $\sigma_0^2 = \int_{-1/2}^{1/2} \phi^2(x)dx = \|\phi(\cdot)\|_2^2$ is a positive constant. To satisfy 2.4.1 (ii) we need the boundedness of this sum. Hence, choose $h_n^{2\beta+1} = 1/n$. This implies

$$2s_n = d(\vartheta_n', \vartheta_n'') = |f_{1n}(x_0)| = \phi(0)\ n^{-\beta/(2\beta+1)}$$

where $\phi(0) \ne 0$ by Definition 2.4.2.

The likelihood ratio in (2.26) for the chosen functions $f_{1n}(x)$ can be written in the form

$$(2.29) \qquad \Lambda(\vartheta_n', \vartheta_n''; X^{(n)}) = \exp\left\{ \Delta_n - \sigma_n^2/2 \right\}$$

where $\Delta_n = \sum_{i=1}^{n} f_{1n}(i/n)\ Y_i$ is zero-mean Gaussian random variable with respect to measure $P_{\vartheta''}$, and its variance is

$$\sigma_n^2 = \sum_{i=1}^{n} f_{1n}^2(i/n) \to \sigma_0^2 \quad \text{as} \quad n \to \infty .$$

Thus, Assumption 2.2.1 is satisfied:

$$P_{\vartheta''}\left(\Lambda(\vartheta',\vartheta'';X^{(n)}) > e^{-\lambda}\right) = P_{\vartheta''}\left(\Delta_n - \sigma_n^2/2 > -\lambda \right) =$$

$$= 1 - \Phi(\sigma_n/2 - \lambda/\sigma_n) \rightarrow 1 - \Phi(\sigma_0/2 - \lambda/\sigma_0)$$

as $n \rightarrow \infty$ where $\Phi(x)=(2\pi)^{-1/2}\int_{-\infty}^{x} \exp(-t^2/2)dt$ is the standard Gaussian c.d.f. □

2.4.4. EXAMPLE. Change-point problem, regular grid design.

Let Y_i be Gaussian observations in the change-point problem. This means that Y_i are independent $(1,\sigma^2)$-Gaussian random variables if $i/n\le\vartheta$, and $(0,\sigma^2)$-Gaussian if $i/n>\vartheta$ (see (1.42) in Section 1.9). In this case $\Theta = R^1$; $\Sigma=(h,1-h)$, $0<h<1/2$; the pseudometric is the absolute value $|\vartheta|$. If ϑ' and ϑ'' belong to the same bin, i.e.

$$(i-1)/n < \vartheta' < \vartheta'' < i/n,$$

and $\vartheta''-\vartheta'=1/(2n)$ then $\Lambda(\vartheta',\vartheta'';X^{(n)})=1$ almost surely with respect to $P_{\vartheta''}$. This implies that Proposition 2.2.2 holds with $s_n=(\vartheta''-\vartheta')/2=(1/4)n^{-1}$. Thus, the I-minimax lower bound (2.5) is true, with the I-lower rate of convergence $\psi_n = n^{-1}$. □

2.4.5. EXAMPLE. Change-point problem, random design.

Let the observations be of the form $Y_i = I(X_i\le\vartheta) + \xi_i$, $i=1,\dots,n$, where ξ_i are independent $(0,\sigma^2)$-Gaussian random variables and X_i are independent uniformly distributed in the interval $[0,1]$. Let us take ϑ' and ϑ'' such that $\vartheta''-\vartheta'=1/n$. It is easy to see that

$$P_{\vartheta''}\left(\Lambda(\vartheta',\vartheta'';X^{(n)}) = 1\right) \ge P_{\vartheta''}\left(X_i \notin [\vartheta',\vartheta''] , \ i=1,\dots,n \right) =$$

$$= (1 - 1/n)^n \ge 1/4$$

for all $n\ge2$. Hence, Assumption 2.2.1 holds with $\lambda=0$ and $p=1/4$. As in the previous example, the lower rate of convergence is equal to n^{-1}.

2.5. THE MAIN THEOREM ON LOWER BOUNDS

The lower bounds in Sections 2.2 and 2.3 were based on the comparison of two hypotheses: $\vartheta'=\vartheta'_n$, $\vartheta''=\vartheta''_n$. However, this technique does not work in all the situations. Thus, if we want to adapt the proof given in Example 2.4.1 to the case when the pseudometric is L_2-metric in the interval [0,1] we immediately face problems. Indeed, let $\vartheta = f(\cdot) \in \Sigma(\beta,L)$ and let

$$d(f,g) = \|f-g\|_2 = \Big(\int_0^1 (f(x)-g(x))^2 dx \Big)^{1/2}.$$

The same construction as in Example 2.4.1 leads to the choice $\vartheta''=\vartheta''_n=0$, and $\vartheta'=\vartheta'_n=f_{1n}(\cdot)$ where

$$\Big(\int_0^1 f_{1n}^2(x)\,dx \Big)^{1/2} = O(n^{-1/2})$$

as $n \to \infty$.

Hence, Proposition 2.2.2 gives the lower rate of convergence $\psi_n=s_n=n^{-1/2}$. However, it is not the minimax rate of convergence for this problem as shown in the next section.

What is the reason for this? Unlike finite-dimensional spaces, the space L_2 contains many functions separated from each other by the distance of order $O(n^{-\beta/(2\beta+1)})$ which is the "suspected" minimax rate. The words "many functions" mean that there is an integer $M=M_n$, $M_n\to\infty$ as $n\to\infty$, such that for some functions $f_{jn} \in L_2$, $j = 0,\dots,M$, we have

$$2s_n = \|f_{jn}(\cdot)-f_{kn}(\cdot)\|_2 = (c + o(1))\, n^{-\beta/(2\beta+1)}, \quad j\neq k,$$

where $o(1)\to 0$ as $n\to\infty$ uniformly in $0\le j<k\le M$. Here c is a positive constant. This forces us to look at the subsets of $\Sigma(\beta,L)$ with growing number of elements in order to get lower bounds.

In other words, instead of applying (2.8) we bound the risk from below as follows:

$$\sup_{\theta\in\Sigma} P_\vartheta\Big(\psi_n^{-1}d(\hat\vartheta_n,\vartheta)\ \ge C\Big) \ge \tag{2.30}$$

$$\ge \max_{\vartheta\in\{\vartheta_0,\vartheta_1,\ldots,\vartheta_M\}} P_\vartheta\Big(\psi_n^{-1}d(\hat\vartheta_n,\vartheta)\ \ge C\Big).$$

Here $\vartheta_0,\vartheta_1,\ldots,\vartheta_M$ are some elements of Σ, and $M=M_n\to\infty$. As far as a discrete set of parameters $\{\vartheta_0,\vartheta_1,\ldots,\vartheta_M\}$ is introduced we are in the framework of hypotheses testing rather than estimation. We have to bound from below the error probability of distinguishing between M hypotheses: $\vartheta=\vartheta_0$, $\vartheta=\vartheta_1,\ldots$, $\vartheta=\vartheta_M$. Therefore, it is not surprising that the methods that we use here are closely related to hypotheses testing and information theory.

Now, the problem is to find lower bounds for the right-hand side of (2.30). We give a solution of this problem under the following assumption.

2.5.1. ASSUMPTION. The elements $\vartheta_0,\vartheta_1,\ldots,\vartheta_M \in \Sigma\subset\Theta$, $M\ge1$, are such that:

(a) for all j,k , $0 \le j < k \le M$, inequality holds

$$d(\vartheta_j,\vartheta_k) \ \ge\ 2s_n \ >\ 0\ ; \tag{2.31}$$

(b) the likelihood ratios have the form

$$\Lambda(\vartheta_0,\vartheta_j;X^{(n)}) \ =\ \exp\Big\{\Delta_j - \lambda_j \log M\Big\} \tag{2.32}$$

where $\lambda_j = \lambda_j^{(n)}$ satisfy the condition $0 < \lambda_j < \lambda < 1$, and the random variables $\Delta_j=\Delta_j^{(n)}$ are positive with positive probability:

$$P_j\Big(\Delta_j \ \ge 0\Big) \ge p_* > 0, \quad j = 1,\ldots,M. \tag{2.33}$$

Here λ and p_* are independent of j and n.

2.5.2. REMARK. In (2.33) and later we write P_j instead of P_{ϑ_j}. Similarly, we write E_j instead of E_{ϑ_j}.

2.5.3. THEOREM. If Assumption 2.5.1 holds then for all estimators $\hat{\vartheta}_n$

$$\max_{0 \le j \le M} P_j\left(d(\hat{\vartheta}_n, \vartheta_j) \ge s_n \right) \ge p_*/2. \tag{2.34}$$

<u>Proof</u> . Assume on the contrary that (2.34) is not true. It means that there exists some estimator ϑ_n^* such that

$$\max_{0 \le j \le M} P_j\left(d(\vartheta_n^*, \vartheta_j) \ge s_n \right) < p_*/2.$$

In particular,

$$P_0\left(d(\vartheta_n^*, \vartheta_0) \ge s_n \right) < p_*/2, \tag{2.35}$$

$$\min_{1 \le j \le M} P_j\left(d(\vartheta_n^*, \vartheta_j) < s_n \right) > 1-p_*/2. \tag{2.36}$$

This together with (2.33) implies

$$P_j\left((d(\vartheta_n^*, \vartheta_j) < s_n) \cap (\Delta_j \ge 0) \right) \ge p_*/2, \; j=1,\dots,M. \tag{2.37}$$

Indeed,

$$P_j\Big((d(\vartheta_n^*, \vartheta_j) < s_n) \cap (\Delta_j \ge 0)\Big) \ge P_j\Big(d(\vartheta_n^*, \vartheta_j) < s_n\Big) +$$

$$+ P_j\left(\Delta_j \ge 0 \right) - 1 \ge (1 - p_*/2) + p_* - 1 \ge p_*/2.$$

As a consequence of (2.31) and of the triangle inequality for the pseudometric d the following property holds: if ϑ_n^* is in "s_n-neighborhood" of a fixed element ϑ_j, $j = 1,\dots,M$, then it is outside of "s_n-neighborhood" of ϑ_0, i.e.

$$\bigcup_{j=1}^{M} \left\{ d(\vartheta_n^*, \vartheta_j) < s_n \right\} \subseteq \left\{ d(\vartheta_n^*, \vartheta_0) \ge s_n \right\}. \tag{2.38}$$

By the same reason the random events $\left\{ d(\vartheta_n^*, \vartheta_j) < s_n \right\}$ are disjoint for different j, $j = 1,\dots,M$. Hence,

(2.39) $$P_0\Big(d(\vartheta_n^*,\vartheta_0) \ge s_n \Big) \ge P_0\Big(\bigcup_{j=1}^{M} \big\{ d(\vartheta_n^*,\vartheta_j) < s_n \big\} \Big) =$$

$$= \sum_{j=1}^{M} P_0\Big(d(\vartheta_n^*,\vartheta_j) < s_n\Big) = \sum_{j=1}^{M} E_j\Big[\Lambda(\vartheta_0,\vartheta_j;X^{(n)}) I(d(\vartheta_n^*,\vartheta_j) < s_n)\Big] =$$

$$= \sum_{j=1}^{M} \exp(-\lambda_j \log M)\, E_j\Big[\exp(\Delta_j)\, I(d(\vartheta_n^*,\vartheta_j) < s_n)\Big] \ge$$

$$\ge M^{-\lambda} \sum_{j=1}^{M} E_j\Big[\exp(\Delta_j)\, I(d(\vartheta_n^*,\vartheta_j) < s_n)\, I(\Delta_j \ge 0) \Big] \ge$$

$$\ge M^{-\lambda} \sum_{j=1}^{M} P_j\Big((d(\vartheta_n^*,\vartheta_j) < s_n) \cap (\Delta_j \ge 0)\Big)$$

where assumption (2.32) has been used. Applying inequality (2.37) to each probability in the last sum we finally get

(2.40) $$P_0\Big(d(\vartheta_n^*,\vartheta_0) \ge s_n \Big) \ge \tfrac{1}{2}p_* M^{1-\lambda} \ge \tfrac{1}{2}p_* .$$

This contradicts (2.35) and proves the theorem. □

2.6. ASSOUAD's LEMMA

To use the main Theorem 2.5.3. directly one needs to construct a growing number of elements which are functions if the space Θ is a functional space. Sometimes it is not a hard work to find the functions satisfying Assumption 2.5.1. The examples will be given in Section 2.7. Unfortunately, the straightforward verification of conditions (2.31) - (2.33) is not easy. To avoid the possible hurdles we present here a method suggested by Bretagnolle and Huber (1979) and in more general form by Assouad (1983). Following Devroye and Györfi (1985) we call it Assouad's lemma. It is not a universal device but it works in many situations connected with nonparametric estimation.

In this section the S-lower rates of convergence are analyzed. They are related to the I-lower rates in the following way.

2.6.1. PROPOSITION. If ψ_n is I-lower rate of convergence for the set Σ in pseudometric $d(\cdot,\cdot)$ then it is S-lower rate of convergence for the set Σ in pseudometric $d(\cdot,\cdot)$.

Proof. If

$$\sup_{\vartheta\in\Sigma} P_\vartheta\left(\psi_n^{-1} d(\hat{\vartheta}_n,\vartheta) \ge C \right) \ge p_0 \tag{2.41}$$

where $p_0>0$ then by Chebyshev's inequality

$$\sup_{\vartheta\in\Sigma} E_\vartheta\left(\psi_n^{-2} d(\hat{\vartheta}_n,\vartheta)^2 \right) \ge C^2 p_0. \quad \square \tag{2.42}$$

The converse statement is not true. The S-lower bounds are weaker than I-lower bounds.

We prove Assouad's lemma for the case of $\vartheta=f(\cdot)\in\Sigma(\beta,L)$. (This does not necessarily mean that the nonparametric regression model is considered.) Assume that the pseudometric d is the L_2-metric.

2.6.2. COROLLARY. If Assumption 2.2.1 is satisfied and $d(\vartheta',\vartheta'')=2s_n>0$ then for any estimator $\hat{\vartheta}_n$

$$E_{\vartheta'}\left(s_n^{-2} d^2(\hat{\vartheta}_n,\vartheta') \right) + E_{\vartheta''}\left(s_n^{-2} d^2(\hat{\vartheta}_n,\vartheta'') \right) \ge \tfrac{1}{2}p\, e^{-\lambda}. \tag{2.43}$$

Proof. Denote $\Sigma = \{\vartheta',\vartheta''\}$. By Proposition 2.2.2 we have

$$\sup_{\vartheta\in\Sigma} P_\vartheta\left(d(\hat{\vartheta}_n,\vartheta) \ge s_n \right) \ge \tfrac{1}{2}p\, e^{-\lambda}.$$

This coincides with (2.41) if we put $\psi_n=s_n$, $C=1$, $p_0=\frac{1}{2}p\, e^{-\lambda}$. Thus, (2.42) holds which in this case has the form

$$\max_{\vartheta=\vartheta',\vartheta''} E_\vartheta\left(s_n^{-2} d^2(\hat{\vartheta}_n,\vartheta) \right) \ge \tfrac{1}{2}p\, e^{-\lambda}.$$

This entails (2.43) since the maximum of two numbers does not exceed their sum. □

2.6.3. ASSUMPTION . There exist non-overlapping subsets V_j of the interval [0,1], and functions $\vartheta_j = f_{jn}(\cdot) \in \Sigma(\beta, L)$, $j=1,\dots,M$, such that

(a) $f_{jn}(x)=0$ if $x \notin V_j$, $j=1,\dots,M$;

(b) $\|f_{jn}(\cdot)\|_2 = 2s_n > 0$ for all $j = 1,\dots,M$;

(c) the likelihood ratios $\Lambda(\vartheta_j, \vartheta_0; X^{(n)})$ satisfy

$$P_0\Big(\Lambda(\vartheta_j, \vartheta_0; X^{(n)}) > e^{-\lambda} \Big) \geq p$$

where positive λ and p are independent of n (here $\vartheta_0 = 0$, i.e. ϑ_0 is the function which equals to 0 everywhere in [0,1], and $P_0 = P_{\vartheta_0}$, see Remark 2.5.2);

(d) if for some functions ϑ' and ϑ'' we have

$$\vartheta' \mathbf{1}_{\overline{V}_j} = \vartheta'' \mathbf{1}_{\overline{V}_j}$$

$$\vartheta' \mathbf{1}_{V_j} = \vartheta_j \mathbf{1}_{V_j} = f_{jn}(\cdot),$$

$$\vartheta'' \mathbf{1}_{V_j} = \vartheta_0 \mathbf{1}_{V_j} = 0$$

($\mathbf{1}_A$ is the indicator function of the set A) then

$$P_{\vartheta''}\Big(\Lambda(\vartheta', \vartheta''; X^{(n)}) > e^{-\lambda} \Big) = P_0\Big(\Lambda(\vartheta_j, \vartheta_0; X^{(n)}) > e^{-\lambda} \Big)$$

The following theorem is a version of the result by Assouad (1983).

2.6.4. THEOREM (Assouad's lemma). If Assumption 2.6.3 is satisfied then for any estimator $\hat{\vartheta}_n$ the following minimax lower bound holds

$$\sup_{\vartheta \in \Sigma(\beta, L)} E_\vartheta\Big((Ms_n^2)^{-1} \|\hat{\vartheta}_n - \vartheta\|_2^2 \Big) \geq \frac{1}{4}\, p\, e^{-\lambda} . \tag{2.44}$$

Proof. Consider the set $\Omega = \Omega_M$ of all binary sequences of length M, i.e. $\bar{\omega} \in \Omega$ if $\bar{\omega} = (\omega_1, \dots, \omega_M)$ where $\omega_j \in \{0,1\}$. Note that

$\mathrm{card}(\Omega) = 2^M$. For each set $\bar\omega = (\omega_1,\ldots,\omega_M) \in \Omega$ define the function

$$f(x,\bar\omega) = \sum_{j=1}^{M} \omega_j f_{jn}(x).$$

It is obvious that

$$(2.45)\qquad \sup_{\vartheta\in\Sigma(\beta,L)} E_\vartheta\Big((Ms_n^2)^{-1}\|\hat\vartheta_n - \vartheta\|_2^2 \Big) \ge \max_{\bar\omega\in\Omega} R(\hat\vartheta_n, f(\cdot,\bar\omega),\psi_n)$$

where $\psi_n = M^{1/2}s_n$ is the rate of convergence, and

$$R(\hat\vartheta_n, f(\cdot,\bar\omega),\psi_n) = E_{f(\cdot,\bar\omega)}\Big((Ms_n^2)^{-1}\|\hat\vartheta_n - f(\cdot,\omega)\|_2^2 \Big).$$

Since the maximum is not less than the average over $\bar\omega\in\Omega$ we get

$$(2.46)\qquad \max_{\bar\omega\in\Omega} R(\hat\vartheta_n, f(\cdot,\bar\omega),\psi_n) \ge 2^{-M} \sum_{\bar\omega\in\Omega} R(\hat\vartheta_n, f(\cdot,\bar\omega),\psi_n).$$

The sets V_j are disjoint, hence, denoting $\hat f_n(\cdot) = \hat\vartheta_n$ we obtain

$$(2.47)\qquad 2^{-M}\sum_{\bar\omega\in\Omega} R(\hat\vartheta_n, f(\cdot,\bar\omega),\psi_n) =$$

$$= 2^{-M}\sum_{\bar\omega\in\Omega}\Big[M^{-1}\sum_{j=1}^{M} E_{f(\cdot,\bar\omega)} \int_{V_j} s_n^{-2}|\hat f_n(x) - f(x,\bar\omega)|^2 dx\Big] =$$

$$= 2^{-M} M^{-1}\sum_{j=1}^{M}\Big[\sum_{\bar\omega\in\Omega} E_{f(\cdot,\bar\omega)} \int_{V_j} s_n^{-2}|\hat f_n(x) - f(x,\bar\omega)|^2 dx\Big].$$

The sum over $\bar\omega$ in (2.47) is in fact the multiple sum :

$$\sum_{\bar\omega\in\Omega} = \sum_{\omega_1=0,1}\cdots\sum_{\omega_M=0,1}.$$

To sum over $\bar\omega_j$ explicitly we introduce for each $\bar\omega \in \Omega$ the notation :

$$\bar{\omega}_{j0} = (\omega_1, \dots, \omega_{j-1}, 0, \omega_{j+1}, \dots, \omega_M),$$

$$\bar{\omega}_{j1} = (\omega_1, \dots, \omega_{j-1}, 1, \omega_{j+1}, \dots, \omega_M).$$

Then the last sum over $\bar{\omega}\in\Omega$ in (2.47) may be expressed as follows

$$(2.48) \qquad \sum{}' \Big[E_{f(\cdot,\bar{\omega}_{j0})} \int_{V_j} s_n^{-2} |\hat{f}_n(x) - f(x,\bar{\omega}_{j0})|^2 dx + E_{f(\cdot,\bar{\omega}_{j1})} \int_{V_j} s_n^{-2} |\hat{f}_n(x) - f(x,\bar{\omega}_{j1})|^2 dx \Big]$$

where

$$\sum{}' = \sum_{\omega_1=0,1} \cdots \sum_{\omega_{j-1}=0,1} \; \sum_{\omega_{j+1}=0,1} \cdots \sum_{\omega_M=0,1}$$

Due to the points (b), (c), (d) of Assumption 2.6.3, for the functions $\vartheta''=f(\cdot,\bar{\omega}_{j0})$ and $\vartheta'=f(\cdot,\bar{\omega}_{j1})$ we have

$$d(\vartheta',\vartheta'') = \|f_{jn}(\cdot)\|_2 = 2s_n > 0,$$

$$P_{\vartheta''}\Big(\Lambda(\vartheta',\vartheta'';X^{(n)}) > e^{-\lambda} \Big) = P_0\Big(\Lambda(\vartheta_j,\vartheta_0;X^{(n)}) > e^{-\lambda} \Big) \ge p.$$

This together with Corollary 2.6.2. guarantees the inequality

$$(2.49) \qquad E_{f(\cdot,\bar{\omega}_{j0})} \int_{V_j} s_n^{-2} |\hat{f}_n(x) - f(x,\bar{\omega}_{j0})|^2 dx + E_{f(\cdot,\bar{\omega}_{j1})} \int_{V_j} s_n^{-2} |\hat{f}_n(x) - f(x,\bar{\omega}_{j1})|^2 dx \ge p_0$$

where $p_0 = p\, e^{-\lambda}/2$.

By substitution of (2.49) and (2.48) into (2.47), we get

$$2^{-M} \sum_{\bar{\omega}\in\Omega} R(\hat{\vartheta}_n, f(\cdot,\bar{\omega}), \psi_n) \ge$$

$$\ge 2^{-M} M^{-1} \sum_{j=1}^{M} \sum{}' p_0 = p_0/2.$$

This completes the proof. □

2.6.5. REMARK. All the results of this section are valid if the L_2-metric is substituted by L_q-metric with some q, $1 \le q < +\infty$, and the loss function is $w(u) = |u|^q$. We say that Assumption 2.6.3 holds for L_q-metric if 2.6.3 (a), (c), (d) hold, and Assumption 2.6.3 (b) is satisfied for the L_q-metric, i.e.

$$s_n = \frac{1}{2}\left(\int |f_{jn}(x)|^q dx \right)^{1/q} > 0, \; j = 1, \ldots, M,$$

If Assumption 2.6.3 holds for L_q-metric then the lower bound in L_q with the rate $\psi_n = M^{1/q} s_n$ is as in (2.44). In particular, for $q=1$ we get the A-lower bound with $\psi_n = M s_n$.

As a conclusion, we summarize the main idea of Assouad's lemma which is now clear from the proof: the problem of proving the lower bound is reduced to M independent problems of distinguishing between pairs of hypotheses.

2.7. EXAMPLES: UNIFORM AND INTEGRAL METRICS

The purpose of this section is to apply the main theorem and Assouad's lemma to nonparametric regression problem in the case of regular grid design

(2.50) $$Y_i = f(i/n) + \xi_i, \quad i = 1,\dots,n,$$

or random design

(2.51) $$Y_i = f(X_i) + \xi_i, \quad i = 1,\dots,n,$$

where ξ_i are independent standard Gaussian errors, $f\in\Sigma(\beta,L)$. Assumptions on the random design points $X_1,\dots,X_n$ are discussed below. Suppose that $\vartheta = f(\cdot)$, and the estimation error is measured by one of two metrics: the uniform metric (C-metric), i.e.

$$d(f,g) = \|f-g\|_\infty = \max_{0\le x\le 1} |f(x)-g(x)|,$$

and L_2-metric defined by the norm $\|\cdot\|_2$.

In the examples of this section we use the following construction. Let $m=m_n>1$ be an integer, and let $\delta_n=1/m_n$. Define the intervals

(2.52) $$U_1=[0,u_1),\ U_2=[u_1,u_2),\dots,\ U_m=[u_{m-1},1]$$

where $u_j=j\delta_n$, $j=0,1,\dots,m$. Let b_j be the center of the interval U_j, and let ϕ be the basic function for $\Sigma(\beta,L)$ defined in Section 2.4.2.

Introduce the functions

(2.53) $$f_{jn}(x) = \delta_n^\beta\,\phi(\delta_n^{-1}(x-b_j)), \quad j = 1,\dots,m,$$

By Proposition 2.4.3 $f_{jn}\in\Sigma(\beta,L)$, $j=1,\dots,m$. The supports of f_{jn} and f_{kn} are non-overlapping if $j\ne k$.

2.7.1. EXAMPLE. C-metric, regular grid design. Let us prove that the I-minimax rate of convergence for nonparametric regression problem (2.50) in C-metric is equal to $(\log n/n)^{\beta/(2\beta+1)}$.

We apply Theorem 2.5.3. Let

$$m=[(n/\log n)^{1/(2\beta+1)}], \quad \delta_n=[(n/\log n)^{1/(2\beta+1)}]^{-1}, \quad M=m.$$

Define the functions

$$\vartheta_j = f_{jn}(\cdot), \quad j = 1,\dots,M,$$

$$\vartheta_0 = 0.$$

The supports of the functions $f_{jn}(x)$ are non-overlapping for different j. Thus, inequalities (2.31) hold with $s_n = \phi_* \delta_n^\beta/2$ where $\phi_* = \max_x \phi(x)$. Indeed,

$$d(\vartheta_j,\vartheta_k) = \|f_{jn}(\cdot) - f_{kn}(\cdot)\|_\infty = \phi_* \delta_n^\beta = 2s_n,$$

$$d(\vartheta_j,\vartheta_0) = \phi_* \delta_n^\beta = 2s_n.$$

The likelihood ratio can be written as

$$\Lambda(\vartheta_0,\vartheta_j;X^{(n)}) = \exp\left\{ \Delta_j - (1/2) \sum_{i=1}^n f_{jn}^2(i/n) \right\} \tag{2.54}$$

where

$$\Delta_j = -\sum_{i=1}^n f_{jn}(i/n) \, (Y_i - f_{jn}(i/n))$$

is zero-mean Gaussian random variable (under P_j) for each j. The variance of Δ_j is

$$\sum_{i=1}^n f_{jn}^2(i/n) = n\delta_n^{2\beta+1} (\|\phi(\cdot)\|_2^2 + o(1)) = \log n \, (\|\phi(\cdot)\|_2^2 + o(1)),$$

$$n \to \infty.$$

Since

$$\log M = (2\beta+1)^{-1} \log n \, (1 + o(1))$$

we have

$$\Lambda(\vartheta_0,\vartheta_j;X^{(n)}) = \tag{2.55}$$

$$= \exp\left\{ \Delta_j - (1/2)(2\beta+1) \log M \left(\|\phi(\cdot)\|_2^2 + o(1)\right)\right\}.$$

If the basic function $\phi(x)$ is chosen to have a sufficiently small L_2-norm, so that $\lambda_j = (1/2)(2\beta+1)(\|\phi(\cdot)\|_2^2+o(1)) < 1$ then (2.32) is valid. Finally, inequality (2.33) is satisfied since the symmetry of the Gaussian distribution entails

$$P_j\left(\Delta_j \geq 0 \right) = 1/2, \; j=1,\ldots,M,$$

when Δ_j is nondegenerate ($\Delta_j \neq 0$ a.s.), and in the degenerate case this probability equals to 1. Thus, Assumption 2.5.1 is holds. Now, it follows from (2.30), (2.34) that $\psi_n = (\log n/n)^{\beta/(2\beta+1)}$ is the I-lower rate of convergence for $\Sigma(\beta,L)$ in C-metric. To prove that this is also I-minimax rate choose $\vartheta_n^* = \hat{f}_n(\cdot)$ where $\hat{f}_n(\cdot)$ is the piecewise-polynomial estimator satisfying assumptions of Theorem 1.8.1. It follows from (1.41) that this estimator satisfies (2.6), and thus it is I-optimal estimator for $\Sigma(\beta,L)$ in C-metric. □

2.7.2. EXAMPLE. C-metric, random design. Let the observations

$$X^{(n)} = ((X_1,Y_1),\ldots,(X_n,Y_n))$$

be defined by regression model (2.51) , and let the random design points $X_1,\ldots,X_n$ be independent and uniformly distributed in the interval [0,1].

As in Example 2.7.1 we apply Theorem 2.5.3. Take

$$m=[(n/\log n)^{1/(2\beta+1)}], \quad \delta_n=[(n/\log n)^{1/(2\beta+1)}]^{-1}, \quad M=m.$$

Define the functions ϑ_j, $j = 0,\ldots,M$, as in Example 2.7.1. The likelihood ratio under P_j is

$$\Lambda(\vartheta_0,\vartheta_j;X^{(n)}) = \tag{2.56}$$

$$= \exp\left\{ -\sum_{i=1}^{n} f_{jn}(X_i)\,(Y_i - f_{jn}(X_i)) - (1/2)\sum_{i=1}^{n} f_{jn}^2(X_i) \right\} =$$

$$= \exp\left\{ - \sum_{i=1}^{n} f_{jn}(X_i)\, \xi_i \;-\; (1/2)\delta_n^{2\beta} \sum_{i=1}^{n} \tilde{\phi}_{jn}(X_i) \right\},\quad j = 1,\ldots,M,$$

where $\tilde{\phi}_{jn}(X_i) = \phi^2(\delta_n^{-1}(X_i - b_j))$, $i=1,\ldots,n$, are independent random variables for each j fixed. Note that since X_i's are uniformly distributed in $[0,1]$ the distribution of $\tilde{\phi}_{jn}(X_i)$ does not depend on j and i. The expectation of each variable $\tilde{\phi}_{jn}(X_i)$ is

$$E(\tilde{\phi}_{jn}(X_1)) = \int \phi^2(\delta_n^{-1}(u - b_j))du = \delta_n \|\phi(\cdot)\|_2^2 ,$$

and its variance is

$$\mathrm{Var}(\tilde{\phi}_{jn}(X_1)) = \int \phi^4(\delta_n^{-1}(u - b_j))du - (\delta_n \|\phi(\cdot)\|_2^2)^2 =$$

$$= \delta_n \int \phi^4(u)du\ (1+o(1)),\quad n\to\infty.$$

By the central limit theorem we have that the distributions of the sums

$$S_{jn} = (n\delta_n)^{-1/2} \sum_{i=1}^{n} \left(\tilde{\phi}_{jn}(X_i) - \delta_n \|\phi(\cdot)\|_2^2 \right).$$

converge to the standard Gaussian distribution, in particular,

$$P(S_{jn} \le 0) = P(S_{1n} \le 0) \to \frac{1}{2} \quad \text{as } n\to\infty.$$

This entails that, uniformly in j,

$$P(S_{jn} \le 0) \ge 1/4 \tag{2.57}$$

for n large. If we introduce the random variable

$$\Delta_j = - \sum_{i=1}^{n} f_{jn}(X_i)\, \xi_i - \frac{1}{2} \delta_n^{2\beta} S_{jn} ,$$

then the likelihood ratio $\Lambda(\vartheta_0, \vartheta_j; X^{(n)})$ can be rewritten as follows

$$\Lambda(\vartheta_0, \vartheta_j; X^{(n)}) = \exp\left\{ \Delta_j - \frac{1}{2} n \delta_n^{2\beta+1} \|\phi(\cdot)\|_2^2 \right\} = \tag{2.58}$$

$$= \exp\left\{ \Delta_j - \frac{1}{2}(\|\phi(\cdot)\|_2^2 + o(1))\log n \right\} .$$

Similarly to (2.55) the second summand in the last expression satisfies (2.32) with $\lambda_j = (1/2)(2\beta+1)(\|\phi(\cdot)\|_2^2+o(1))<1$. To verify the inequality (2.33) one can note that

$$P_j\left(\Delta_j \ge 0 \right) \ge P_j\left(S_{jn} \le 0 ; \sum_{i=1}^{n} f_{jn}(X_i)\, \xi_i \le 0 \right) =$$

$$= E_j\left[I(S_{jn} \le 0)\, P_j\left(\sum_{i=1}^{n} f_{jn}(X_i)\, \xi_i \le 0 \mid X_1,\dots,X_n \right) \right] .$$

If the values $X_1,\dots,X_n$ are fixed then the conditional distribution of the sum $\sum_{i=1}^{n} f_{jn}(X_i)\xi_i$ is zero-mean Gaussian, possibly degenerate.Hence,

$$P_j\left(\sum_{i=1}^{n} f_{jn}(X_i)\, \xi_i \le 0 \mid X_1,\dots,X_n \right) = \frac{1}{2} \quad \text{(a.s.)},$$

or in the degenerate case this probability is equal to 1. Using (2.57), we have

$$\text{(2.59)} \qquad P_j\left(\Delta_j \ge 0 \right) \ge \frac{1}{2}\, P\left(S_{jn} \le 0 \right) \ge \frac{1}{8}$$

if n is large enough. Thus, Assumption 2.5.1 is satisfied and, consequently, $\psi_n=(\log n/n)^{\beta/(2\beta+1)}$ is the I-lower rate of convergence. In view of Remark 1.8.2 this sequence is the I-minimax rate of convergence.

2.7.3. EXAMPLE. L_2-metric, regular grid design. Now let us return to the example considered at the beginning of Section 2.5. Let

$$\vartheta = f(\cdot) \in \Sigma(\beta,L) \quad \text{and} \quad d(f,g) = \|f-g\|_2 .$$

If the observations $X^{(n)} = (Y_1,\dots,Y_n)$ satisfy (2.50) then the

S-minimax rate of convergence proves to be $\psi_n = n^{-\beta/(2\beta+1)}$. Indeed, let

$$m=[n^{1/(2\beta+1)}], \quad \delta_n=[n^{1/(2\beta+1)}]^{-1}, \quad M=m,$$

and let $\vartheta_0=0$, and the functions $\vartheta_j = f_{jn}(\cdot)$, $j=1,\dots,M$, be defined by (2.53). In this example we would like to use Assouad's lemma, so we have to verify Assumption 2.6.3. It is easy to see that Assumptions 2.6.3(a) and 2.6.3(b) hold if we put $V_j = U_j$ and

$$2s_n = \|f_{jn}(\cdot)\|_2 = (\delta_n^{2\beta+1}\|\phi(\cdot)\|_2^2)^{1/2} = (\|\phi(\cdot)\|_2+o(1))\ n^{-1/2}.$$

The likelihood ratio is expressed by (2.26), hence, Assumption 2.6.3(c) is satisfied. The last part (d) of Assumption 2.6.3 is also obvious. Application of Theorem 2.6.4 leads to (2.44) with the following S-lower rate of convergence

$$(2.60) \qquad \psi_n = M^{1/2}s_n = (1/2)\ [n^{1/(2\beta+1)}]^{1/2}(\|\phi(\cdot)\|_2+o(1))n^{-1/2} =$$

$$= \left(\|\phi(\cdot)\|_2/2 + o(1) \right) n^{-\beta/(2\beta+1)},$$

as $n \to \infty$. Comparing (2.44) with Theorem 1.6.8 we find that ψ_n given by (2.60) is the S-minimax rate of convergence. Moreover, the PPE(k) and LPE(k) satisfying the assumptions of Theorem 1.6.8 are optimal estimators for $\Sigma(\beta,L)$ in L_2-metric. □

It is clear from the proof of Assouad's lemma that it is suited well for integral metrics such as L_2-metric. It provides the brief and elegant proof of minimax lower bounds. If we try to use Theorem 2.5.3 directly, we shall also obtain such bounds but in much more tedious way. Nevertheless, we will give now a relevant construction since it is based on the techniques that may be useful in different situations. It has been originally proposed by Ibragimov and Khasminskii (1980a).

Assume that the conditions of Example 2.7.3 are satisfied. Consider the set Ω_m of all possible 2^m binary vectors $\bar{\omega}=(\omega_1,\dots,\omega_m)$, $\omega_\ell\in\{0,1\}$, $\ell=1,\dots,m$. For each vector $\bar{\omega}$ define the function

$$(2.61) \qquad f(x,\bar{\omega}) = \sum_{j=1}^{m} \omega_j f_{jn}(x)$$

where f_{jn} are the functions (2.53) with

$$m=[n^{1/(2\beta+1)}], \quad \delta_n=[n^{1/(2\beta+1)}]^{-1}.$$

Since $f_{jn}\in\Sigma(\beta,L)$ and the supports of f_{jn} are non-overlapping we have $f(\cdot,\bar{\omega})\in\Sigma(\beta,L)$. The total number of functions (2.61) is 2^m. We don't identify all these functions with the hypotheses $\vartheta_1,\ldots,\vartheta_M$ in (2.30). Instead of this we choose the hypotheses from a subset of cardinality $M<2^m$ containing all functions of the form (2.61) which are well separated from each other. To define this subset we use a result from information theory known as the Varshamov-Gilbert bound.

Denote by $\rho_H(\bar{\omega}',\bar{\omega}'')$ the Hamming distance between the vectors $\bar{\omega}'=(\omega_1',\ldots,\omega_m')$ and $\bar{\omega}''=(\omega_1'',\ldots,\omega_m'')$, i.e.

$$\rho_H(\bar{\omega}',\bar{\omega}'') = \text{card}\Big\{\ell : 1\le\ell\le m,\ \omega_\ell' \ne \omega_\ell''\Big\}.$$

Define $\bar{\omega}_0=(0,\ldots,0)$.

2.7.4. LEMMA (The Varshamov-Gilbert bound). There exist $M=[2^{m/8}]$ vectors $\bar{\omega}_1,\ldots,\bar{\omega}_M$, such that $\rho_H(\bar{\omega}_j,\bar{\omega}_k) \ge m/16$, $0\le j<k\le M$.

Proof. Consider the Hamming ball of radius $m/16$ around $\bar{\omega}_0$:

$$B(\bar{\omega}_0) = \Big\{\bar{\omega}:\ \rho_H(\bar{\omega}_0,\bar{\omega})<m/16\Big\}.$$

The number of binary vectors in this ball is

$$\text{card}\, B(\bar{\omega}_0) \le \sum_{k=1}^{[m/16]} \binom{m}{k} = 2^m\tau$$

where $\tau = \sum_{k=1}^{[m/16]} 2^{-m}\binom{m}{k}$. Note that τ is the binomial probability, and it can be estimated directly. The elegant way of estimating τ is based on the Hoeffding's (1963) inequality: for $\tilde{\eta}_1,\ldots\tilde{\eta}_m$, i.i.d. random variables such that $0 \le \tilde{\eta}_i \le 1$, $i=1,\ldots,m$, one has

$$P\left(\frac{1}{m}\sum_{i=1}^{m}(E(\tilde{\eta}_i) - \tilde{\eta}_i) \geq \varepsilon\right) \leq \exp(-2m\varepsilon^2),\ \forall\varepsilon>0.$$

Let $\tilde{\eta}_i$'s are i.i.d. Bernouillian variables, $E(\tilde{\eta}_i)=1/2$. Using the Hoeffding's inequality we get

$$\tau = P\left(\sum_{i=1}^{m}\tilde{\eta}_i \leq [m/16]\right) \leq P\left(\sum_{i=1}^{m}(\tilde{\eta}_i - E(\tilde{\eta}_i)) \leq -3m/8\right) \leq 2^{-m/4}.$$

Hence, $\text{card}\, B(\bar{\omega}_0) \leq 2^{3m/4}$. Consider the set $\Omega_m \backslash B(\bar{\omega}_0)$. The cardinality of this set is greater than 2^{m-1} for m large enough. Assume that Lemma 2.7.4 is not true. Then for any vectors $\bar{\omega}_1,\dots,\bar{\omega}_{M-1},\bar{\omega}$ from $\Omega_m\backslash B(\bar{\omega}_0)$ we have $\rho_H(\bar{\omega}_j,\bar{\omega})<m/16$ at least for one $j\in\{1,\dots,M-1\}$. Thus, the balls $B(\bar{\omega}_j)$ of radius m/16 centered at $\bar{\omega}_j$, $j=1,\dots,M-1$, cover $\Omega_m\backslash B(\bar{\omega}_0)$. Since $\text{card}\, B(\bar{\omega}_j)\leq 2^{3m/4}$ we have

$$2^{m-1} \leq \text{card}\,(\Omega_m\backslash B(\bar{\omega}_0)) \leq \sum_{j=1}^{M-1}\text{card}\, B(\bar{\omega}_j) \leq M\, 2^{3m/4}.$$

This contradiction proves the lemma. □

Now we start to check the Assumption 2.5.1. Put $\vartheta_0=0$, and $\vartheta_j=f(\cdot,\bar{\omega}_j)$, $j=1,\dots,M=[2^{m/8}]$, where $\bar{\omega}_1,\dots,\bar{\omega}_M$ are the binary vectors from Lemma 2.7.4. Inequalities (2.31) in Assumption 2.5.1 hold since

$$\text{(2.62)}\qquad d(\vartheta_j,\vartheta_k) = \|f(\cdot,\bar{\omega}_j) - f(\cdot,\bar{\omega}_k)\|_2 =$$

$$= \left(\rho_H(\bar{\omega}_j,\bar{\omega}_k)\,\delta_n^{2\beta+1}\|\phi(\cdot)\|_2^2\right)^{1/2} \geq 2s_n,\ j\neq k,$$

with

$$\text{(2.63)}\qquad s_n = (1/2)(\delta_n^{2\beta+1}\|\phi(\cdot)\|_2^2\ m/16)^{1/2} =$$

$$= \left(\|\phi(\cdot)\|_2/8 + o(1)\right)\, n^{-\beta/(2\beta+1)}.$$

The likelihood ratio under P_j is

$$\text{(2.64)}\qquad \Lambda(\vartheta_0,\vartheta_j;X^{(n)}) = \exp\left\{\Delta_j - (1/2)\sum_{i=1}^{n} f^2(i/n,\bar{\omega}_j)\right\}$$

where $\Delta_j = -\sum_{i=1}^{n} f(i/n,\bar{\omega}_j)\xi_i$ is zero-mean Gaussian random variable satisfying (2.33), and

$$(2.65)\qquad (1/2)\sum_{i=1}^{n} f^2(i/n,\bar{\omega}_j) \le (1/2)mn\delta_n^{2\beta+1}(\|\phi(\cdot)\|_2^2 + o(1)) \le$$

$$\le (4/\log 2)\log M\left(\|\phi(\cdot)\|_2^2 + o(1)\right) = \lambda_j \log M.$$

If $\|\phi(\cdot)\|_2^2$ is small enough and n is sufficiently large then $\lambda_j \le \lambda < 1$. Relations (2.62)-(2.65) show that Assumption 2.5.1 holds and Theorem 2.5.3 proves the I-lower rate of convergence $\psi_n = n^{-\beta/(2\beta+1)}$.

2.7.5. EXAMPLE. L_2-metric, random design. Here we present another application of the Varšhamov-Gilbert bound. Consider the nonparametric regression model with random design as in Example 2.7.2 but assume that the estimation error is measured in L_2-metric. Put $\vartheta_0 = 0$, and $\vartheta_j = f(\cdot,\bar{\omega}_j)$, $j=1,\dots,M$, where $\bar{\omega}_1,\dots,\bar{\omega}_M$ are the binary vectors from Lemma 2.7.4. Then Assumption 2.5.1(a) is satisfied in view of (2.62). The likelihood ratio has the form

$$(2.66)\qquad \Lambda(\vartheta_0,\vartheta_j;X^{(n)}) = \exp\left\{-\sum_{i=1}^{n} f(X_i,\bar{\omega}_j)\xi_i - (1/2)\sum_{i=1}^{n} f^2(X_i,\bar{\omega}_j)\right\}$$

which is similar to (2.64). The expectation of the second sum in the right-hand side of (2.66) is equal to

$$(1/2)n\,\rho_H(\bar{\omega}_j,\bar{\omega}_0)\,\delta_n^{2\beta+1}\|\phi(\cdot)\|_2^2 \le (1/2)m(\|\phi(\cdot)\|_2^2 + o(1)) =$$

$$= (4/\log 2)\log M\,(\|\phi(\cdot)\|_2^2 + o(1)) = \lambda_j \log M.$$

Note that $\rho_H(\bar{\omega}_j,\bar{\omega}_0)$ is the number of non-zero elements in $\bar{\omega}_j$. The same argument as in Example 2.7.2 and formula (2.59) entail that Assumption 2.5.1 is satisfied. Hence, the S-lower rate of convergence coincides with that for the case of the regular grid design.

2.7.6. MULTIDIMENSIONAL CASE. Let us discuss the multidimensional extensions of nonparametric regression models (2.50) and (2.51). For any $\beta>0$ consider the functions $f=f(x_1,\dots,x_N)$, $f:[0,1]^N \to R^1$ having continuous partial derivatives up to the order $k=\lfloor\beta\rfloor$. Denote by $p_x(z)$ the Taylor polynomial of order k for $f(z)$ at point $x\in[0,1]^N$. Let $\Sigma_N(\beta,L)$, $L>0$, be the class of functions f such that

$$|f(z) - p_x(z)| \le L|z-x|^{\beta}, \ \forall\ x,z\in[0,1]^N,$$

where $|z|$ denotes the Euclidean norm of the vector $z\in[0,1]^N$. Consider the multidimensional nonparametric regression model

$$Y_i = f(X_i) + \xi_i, \quad i=1,\dots,n,$$

where ξ_i are i.i.d. Gaussian errors; X_i are design points in $[0,1]^N$, and $f\in\Sigma_N(\beta,L)$. If $n=n_1^N$ is an integer, then the regular grid design $(X_1,\dots,X_n)$ can be defined as a set of points in $[0,1]^N$ with coordinates proportional to $1/n_1$. The random uniform design is defined similar to one-dimensional case as a set of independent points X_i uniformly distributed in $[0,1]^N$. The minimax rate of convergence in L_2-metric is equal to $n^{-\beta/(2\beta+N)}$ for the class $\Sigma_N(\beta,L)$. The proof of this fact is quite identical to that in Examples 2.7.3 and 2.7.5. The same is true for the minimax rate of convergence in C-metric which is equal to $(\log n/n)^{\beta/(2\beta+N)}$.

2.8. ARBITRARY DESIGN

Deterministic design. In this section we are going to discuss the role of design for the lower bounds in nonparametric regression problem. Let

$$(2.67) \qquad Y_i = f(X_i) + \xi_i, \quad i=1,\dots,n, \quad f\in\Sigma(\beta,L).$$

We assume for simplicity that ξ_i in (2.67) are independent standard Gaussian errors. Let the design points $X_1,\dots,X_n$ be deterministic and known to the statistician. No assumptions on $X_1,\dots,X_n$ are imposed except that $X_i\in[0,1]$.

It is natural to ask whether the minimax rates of convergence obtained in examples of Sections 2.4 and 2.7 are the same under arbitrary design. For some pseudometrics the answer is negative. Indeed, let the pseudometric d be equal to the distance at some given point $x_0 \in [0,1]$: $d(f,g)=|f(x_0)-g(x_0)|$ (Example 2.4.1). If all the design points coincide with x_0, i.e. $X_i=x_0$, $i=1,\ldots,n$, then observations (2.67) have the form

$$Y_i = f(x_0) + \xi_i, \quad i=1,\ldots,n.$$

The problem now is to estimate the location parameter $f(x_0)$. It is well-known that the minimax rate of convergence for this parametric problem equals $n^{-1/2}$ (see e.g. Ibragimov and Khasminskii (1981)). This can be easily verified by the method of Section 2.2 as well. What is more, the routine estimator $\vartheta_n^*=(Y_1+\ldots+Y_n)/n$ is efficient in the following sense :

$$E_{f(\cdot)}\left(n^{1/2}(\vartheta_n^* - f(x_0))\right)^2 = 1 \tag{2.68}$$

for any $f \in \Sigma(\beta,L)$.

This example shows that the minimax rate of convergence may be sensitive to the choice of design points $X_1,\ldots,X_n$. However, it turns out that for L_q-metrics the minimax rates do not depend on the design. Let us prove this for L_2-metric (the extension to general L_q is easy).

Let

$$m=[n^{1/(2\beta+1)}], \quad \delta_n=[n^{1/(2\beta+1)}]^{-1},$$

and let $U_1,\ldots,U_m$ be the intervals (2.52). Assume that n is large enough so that $m > 2$.

2.8.1. LEMMA . Let $\{X_1,\ldots,X_n\}$ be an arbitrary set of points in the interval [0,1]. If $m>2$ then there exist at least $M = [m/2]$ intervals $V_1=U_{\ell_1},\ldots,V_M=U_{\ell_M}$ among $U_1,\ldots,U_m$, each containing no more than $2([n/(m-1)]+1)$ points of the set $\{X_1,\ldots,X_n\}$.

<u>Proof</u>. Let each set of M intervals contain more than $2([n/(m-1)]+1)$ points. Then the total number of points is at least $2M([n/(m-1)]+1) = 2[m/2]([n/(m-1)]+1) > (m-1)(n/(m-1)) = n$. This contradiction proves the lemma. □

2.8.2. PROPOSITION. Let $\{X_1,\dots,X_n\}$ be an arbitrary deterministic design in nonparametric regression (2.67). Let $M=[m/2]$ and let the intervals $V_1=U_{\ell_1},\dots,V_M=U_{\ell_M}$ be defined by Lemma 2.8.1. Then there exist the functions

$$\vartheta_j = f_{jn}(\cdot)\in\Sigma(\beta,L), \quad j = 1,\dots,M,$$

which satisfy Assumption 2.6.3 with $s_n \geq cn^{-1/2}$, where the positive c is independent of n.

Proof. Let $\phi(x)$ be the basic function introduced in Section 2.7.2. Define $\vartheta_j = f_{\ell_j n}(\cdot) = \delta_n^\beta \, \phi(\delta_n^{-1}(\cdot-b_{\ell_j}))$, $j=1,\dots,M=[m/2]$, where b_{ℓ_j} are the centers the of intervals $V_j = U_{\ell_j}$, $j = 1,\dots,M$, and $\vartheta_0 = 0$. Then Assumption 2.6.3(b) holds with

$$2s_n = \|f_{jn}(\cdot)\|_2 = (\delta_n^{2\beta+1}\|\phi(\cdot)\|_2^2)^{1/2} = (\|\phi(\cdot)\|_2+o(1))n^{-1/2}.$$

Hence, only Assumption 2.6.3(c) is to be verified. For any j, $1\leq j\leq M$, the likelihood ratio is

$$\Lambda(\vartheta_j,\vartheta_0;X^{(n)}) = \exp\left\{ \sum_{X_i\in V_j} f_{\ell_j n}(X_i)Y_i - (1/2)\sum_{X_i\in V_j} f^2_{\ell_j n}(X_i) \right\}.$$

Since the intervals V_j are chosen according to Lemma 2.8.1 we get

$$(1/2)\sum_{X_i\in V_j} f^2_{\ell_j n}(X_i) \leq \phi_*^2\delta_n^{2\beta}([n/(m-1)] + 1) \leq$$

$$\leq \phi_*^2\delta^{2\beta}(n/m)(1+o(1)) = 2\phi_*^2 + o(1), \ n\to\infty,$$

where $\phi_* = \max_x \phi(x)$. The same argument as in (2.28) shows that Assumption 2.6.3 is satisfied. □

2.8.3. COROLLARY. In nonparametric regression problem (2.67) with arbitrary deterministic design the S-lower rate of convergence for $\Sigma(\beta,L)$ in the L_2-metric is $\psi_n = n^{-\beta/(2\beta+1)}$.

To prove this corollary it suffices to apply Proposition

2.8.2 and Theorem 2.6.4 (Assouad's lemma).

Using Corollary 2.8.3 and Theorem 1.6.8 we get that $\psi_n=n^{-\beta/(2\beta+1)}$ is the S-minimax rate of convergence, and the estimators of Theorem 1.6.8 are optimal for any deterministic design.

Random design. The rest of this section is devoted to the model (2.67) with random design $\{X_1,\dots,X_n\}$. We assume that the distribution of random vector $(X_1,\dots,X_n)$ is arbitrary in $[0,1]^n$ and that $\{X_1,\dots,X_n\}$ is independent of $\{\xi_1,\dots,\xi_n\}$. If the estimation error is measured in L_2-metric then the S-minimax rate of convergence under random design is the same as under deterministic design. To prove this we cannot use Proposition 2.8.2 directly since the expectation E_ϑ in the squared risk function $R(\hat{\vartheta}_n,\vartheta,\psi_n)=E_\vartheta\left(\psi_n^{-2}\|\hat{\vartheta}_n-\vartheta\|_2^2\right)$ is now taken with respect to both design points and random errors. Thus, a certain penetration into the proof of the Theorem 2.6.4 is required .

2.8.4. THEOREM. Let the random design points $\{X_1,\dots,X_n\}$ in (2.67) be independent of errors $\{\xi_1,\dots,\xi_n\}$. Then the S-lower rate of convergence for $\Sigma(\beta,L)$ in the L_2-metric is $\psi_n=n^{-\beta/(2\beta+1)}$.

Proof. Let

$$m=[n^{1/(2\beta+1)}],\quad \delta_n=[n^{1/(2\beta+1)}]^{-1},$$

and let the functions $f_{\ell n}(x) = \delta_n^\beta\ \phi(\delta_n^{-1}(x-b_\ell))$ be defined for all intervals U_ℓ, $\ell=1,\dots,m$. For an arbitrary vector $\bar{\omega}=(\omega_1,\dots,\omega_m)$ of binary variables we define the function

$$f(x,\bar{\omega}) = \sum_{\ell=1}^{m} \omega_\ell f_{\ell n}(x).$$

Similarly to (2.46), (2.47) we obtain the following inequality for any estimator $\hat{\vartheta}_n=\hat{f}_n(\cdot)$:

$$\sup_{f(\cdot)\in\Sigma(\beta,L)} E_{f(\cdot)}\left((n/m)\|\hat{\vartheta}_n - f(\cdot)\|_2^2 \right) \ge \tag{2.69}$$

$$\geq 2^{-m} m^{-1} \sum_{\ell=1}^{m} E_{f(\cdot,\bar{\omega})}\Big(\sum_{\bar{\omega}\in\Omega} E_{f(\cdot,\bar{\omega})}\Big[\int_{U_\ell} n|\hat{f}_n(x) - f(x,\bar{\omega})|^2 dx | X_1,\dots,X_n\Big]\Big)$$

where $E_{f(\cdot,\bar{\omega})}\Big[\ \cdot\ |X_1,\dots,X_n\Big]$ is the conditional expectation which is identical to the expectation with respect to random errors $\{\xi_1,\dots,\xi_n\}$ due to the independence of design and noise. Let $V_j = U_{\ell_j}$, $j=1,\dots,M=[m/2]$, be the intervals from Lemma 2.8.1. Clearly, these intervals depend on the design points $\{X_1,\dots,X_n\}$. Thus, the right-hand side of (2.69) is bounded from below by

$$2^{-m} m^{-1} \sum_{j=1}^{M} E_{f(\cdot,\bar{\omega})}\Big(\sum_{\bar{\omega}\in\Omega} E_{f(\cdot,\bar{\omega})}\Big[\int_{V_j} n|\hat{f}_n(x) - f(x,\bar{\omega})|^2 dx | X_1,\dots,X_n\Big]\Big)$$

The sum $\sum_{\bar{\omega}\in\Omega}$ over $(\omega_1,\dots,\omega_m)$ may be split into two sums :

$$\sum_{\bar{\omega}\in\Omega} = \sum_{\omega_\ell:\ell\neq\ell_1,\dots,\ell\neq\ell_M} \ \sum_{\omega_\ell:\ell=\ell_1,\dots,\ell=\ell_M} .$$

Numbers of summands in these sums are respectively equal to 2^{m-M} and 2^M. Hence,

$$\sup_{f(\cdot)\in\Sigma(\beta,L)} E_{f(\cdot)}\Big((n/m)\|\hat{\vartheta}_n - f(\cdot)\|_2^2 \Big) \geq \tag{2.70}$$

$$\geq 2^{-(m-M)} (M/m) E_{f(\cdot,\bar{\omega})}\Big(\sum_{\omega_\ell:\ell\neq\ell_1,\dots,\ell\neq\ell_M} S_n(\bar{\omega},X_1,\dots,X_n) \Big),$$

where

$$S_n(\bar{\omega},X_1,\dots,X_n) =$$

$$2^{-M}M^{-1}\sum_{\omega_\ell:\ell=\ell_1,\dots,\ell=\ell_M}\left\{\sum_{j=1}^{M}E_{f(\cdot,\bar\omega)}\left[\left[\int_{V_j}n|\hat f_n(x)-f(x,\bar\omega)|^2dx|X_1,\dots,X_n\right]\right]\right\}.$$

For almost all sets of design points $\{X_1,\dots,X_n\}$ the function $S_n(\bar\omega,X_1,\dots,X_n)$ is proportional to the right-hand side of (2.47). The same argument as in (2.48), (2.49) shows that $S_n(\bar\omega,X_1,\dots,X_n)\ge p_*$ with some positive p_* independent of n and of $\{X_1,\dots,X_n\}$. Note that the condition m >2 entails $M/m \ge 1/3$. This together with (2.70) gives

$$\sup_{f(\cdot)\in\Sigma(\beta,L)} E_{f(\cdot)}\left((n/m)\|\hat\vartheta_n - f(\cdot)\|_2^2 \right) \ge p_*/3. \ \square$$

2.8.5. REMARK. The results of this section can easily be extended to the multidimensional nonparametric regression model (see Example 2.7.6). The S-lower rate of convergence for $\Sigma_N(\beta,L)$ in L_2-metric is thus $n^{-\beta/(2\beta+N)}$ also in the case of arbitrary design. For two-dimensional regression (N=2) this gives the lower rate $n^{-\beta/(2(\beta+1))}$.

CHAPTER 3. THE PROBLEM OF EDGE AND IMAGE ESTIMATION

3.1. INTRODUCTION

One of the main problems in image analysis is reconstruction of pictures from noisy data, or image estimation (see e.g. the books of Pratt (1978), Grenander (1976,1978,1981), Rosenfeld and Kak (1982), Marr (1982), Blake and Zisserman (1987)). There are two special features related to this problem. First, the data arrays are two-dimensional (or multidimensional). Second, the image is usually composed of several regions with rather sharp edges. Within each region the image preserves a certain degree of uniformity while on the boundaries between the regions it has considerable changes. Therefore, it is important to find the boundaries (edges) of such regions. This leads to edge detection or edge estimation problems.

A large variety of methods has been proposed for solving the image and edge estimation problems in different contexts. An overview of them can be found in the above mentioned books. The most popular methods are based on certain parametric image models with large number of parameters and on Bayesian approach to estimation (Haralick (1980), Geman and Geman (1984), Ripley (1988) among others). Some nonparametric methods, however, are also available in image estimation.They are related to penalizing and regularization techniques (Titterington (1985), Torre and Poggio (1986), Shiau, Wahba and Johnson (1986), Mumford and Shah (1989), Girard (1990)).

The edge detection methods mostly have the local character and do not assume any underlying parametric model (Pratt (1978), Marr (1982), Nagao and Matsuyama (1979), Lee (1983), Huang and Tseng (1988), Müller and Song (1992)). The common way of edge detection discussed in these works is based on smoothing by a kernel concentrated on a small interval and with positive and negative values. Such kernels mimic differentiation. However, because of the linear character of differentiation operators such methods do not achieve the optimal rates of convergence if the edge is really a discontinuity (cf. Section 6.4).

The main purpose of this book is to develop the tools of comparing different image and edge estimators on the common scale. For this we apply the minimax approach. In Chapters 1 and 2 we have shown that the minimax approach is suited well for comparison of estimators in nonparametric regression and change-point problems. Here we introduce a minimax nonparametric framework for image and edge estimation based on similar ideas. The main results of this chapter and of Chapters 4-9 can be found in Korostelev (1988, 1991), Korostelev and Tsybakov (1989, 1990, 1991, 1992a,b,c), Tsybakov (1989, 1991, 1992), Mammen and Tsybakov (1992). The interesting related work is due to Rudemo, Skovgaard and Stryhn (1990), Rudemo and Stryhn (1991) and Carlstein and Krishnamoorthy (1992) who apply nonparametric smoothing techniques to edge estimation models somewhat different from ours.

Let us introduce the statistical model of noisy image.

Consider the sample of observations

$$Y_i = f(X_i) + \xi_i, \quad i=1,\dots,n, \tag{3.1}$$

where f is an unknown function of two-dimensional argument, X_i's belong to the unit square $K=[0,1]\times[0,1]$ and ξ_i are random variables. The function f is called <u>image</u>. The <u>design points</u> X_i can be fixed or random. In this book it is assumed that ξ_i are i.i.d. and that $(\xi_1,\dots,\xi_n)$ is independent of the design points. Assumptions on the design and on the distribution of noise will be given later. Denote

$$X^{(n)} = ((X_1,Y_1), \dots, (X_n,Y_n)) .$$

We associate an image with a domain G inside K. This means that K is divided into two parts: the <u>domain</u> G where the values of f are large and the <u>background</u> $\overline{G}=K\backslash G$ where the values are small. We assume that f has a jump discontinuity at the boundary $\Gamma=\partial G$ of the domain G. The curve Γ is called <u>edge</u>.

The following two image models are considered.

3.1.1. BINARY IMAGE.

$$f(x) = I(x\in G) = \begin{cases} 1 & \text{if } x\in G, \\ 0 & \text{if } x\in \overline{G}. \end{cases}$$

Reconstruction of binary images reduces to estimation of the domain G from observations (3.1), or, equivalently, to estimation of the edge Γ.

3.1.2. GREY-SCALE IMAGES.

$$f(x) = \begin{cases} f_1(x) \text{ if } x \in G, \\ f_0(x) \text{ if } x \in \bar{G}. \end{cases}$$

The values of f_0 and f_1 should be separated in a certain way. For example, it is sufficient to assume that

$$f_0(x) \le a < b \le f_1(x)$$

with some known or unknown constants a and b.

In the following we mean by <u>image estimation</u> the estimation of grey-scale images f from observations (3.1), and by <u>edge estimation</u> the estimation of G and Γ for binary or grey-scale images.

Clearly, the problems of image and edge estimation can be stated also in N-dimensional case (N>2). Then the image is defined as a function on N-dimensional cube $K=[0,1]^N$, and the design points X_i belong to $[0,1]^N$. We study the N-dimensional problems later in Chapter 5. Here and in Chapter 4 only the two-dimensional images are considered.

In fact, the case N=2 is most interesting for the majority of traditional image analysis applications. Also, as we show later, the main mathematical features of image and edge estimation problems appear already in two-dimensional setup. The extension to higher dimensions is usually straightforward. In the following we assume that N=2, with all exclusions specially marked.

One of the examples showing that high-dimensional binary images are worth studying is related to neural networks. Consider the simplest neural network (the two-layer feedforward network). It contains N input neurons which send signals x_j, j=1,..,N, towards the output neuron. The well-known model of Rosenblatt's perceptron (see Minsky and Papert (1969)) assumes that the output neuron does not react on these signals unless some function $g(x_1,\ldots,x_N)$ (e.g. the linear combination $\alpha_1 x_1+\ldots+\alpha_N x_N$ where

$\alpha_1,\ldots,\alpha_N$ are some coefficients) exceeds some threshold α. If the threshold is overshot the output neuron "fires". Formally it is expressed as follows: the signal sent by the output neuron is the indicator function $I(g(\alpha_1,\ldots,\alpha_N)\geq\alpha)$. In general, the output neuron fires if $X=(\alpha_1,\ldots\alpha_N)$ belongs to some unknown set G. With random noise ξ_i and repeated measurements the model has the form

$$Y_i = I(X_i\in G) + \xi_i, \quad i=1,\ldots,n,$$

where Y_i is the output value, and $X_i=(\alpha_{1i},\ldots,\alpha_{Ni})$ is the input vector for ith observation.

3.2. ASSUMPTIONS AND NOTATION

Let us introduce some basic definitions that will be used throughout the book. We start with definitions of various types of design.

3.2.1. ASSUMPTIONS ON THE DESIGN. In Section 1.9 we defined four types of one-dimensional designs. Here we extend these definitions to the two-dimensional case keeping the same notation.

Note that in most of the work on image analysis the choice of design is not discussed. It is usually taken for granted that the design points are on the regular grid, i.e. that the following assumption holds.

(A1) Regular grid. Let $n_1=n^{1/2}$ be an integer. Define

$$X_i=(k/n_1,\ell/n_1)$$

where $i=(k-1)n_1+\ell$, $k,\ell=1,\ldots,n_1$. The set $\{X_1,\ldots,X_n\}$ is called regular grid.

Sometimes it is assumed that an image or some part of it are blurred. This could be expressed by the following assumption on the design.

(A2) Random shift of the regular grid. Let $n_1=n^{1/2}$ be an integer. Define

$$X_i=(k/n_1-\eta_1,\ell/n_1-\eta_2)$$

where $i=(k-1)n_1+\ell$, $k,\ell=1,\ldots,n_1$, and η_1, η_2 are independent random variables uniformly distributed in $[0,1/n_1]$.

Designs (A1), (A2) are called regular. It will be shown later that regular designs impose strong restrictions on the convergence rates of image and edge estimators. If the edge is smooth enough then the minimax convergence rates cannot be achieved with regular designs. Therefore, we have to use some other designs under which the minimax rates are possible. These are, for example:

(B1) Random uniform design. The design points X_i are independent random points uniformly distributed in K.

(B2) Jittered regular grid. Let $n_1=n^{1/2}$ be an integer. Define

$$X_i=\eta_{k\ell} \quad \text{for } i=(k-1)n_1+\ell,$$

where $\eta_{k\ell}$ are independent random points uniformly distributed in the small squares $[(k-1)/n_1, k/n_1]\times[(\ell-1)/n_1, \ell/n_1]$, $k,\ell=1,\ldots,n_1$.

Later we call (B1) and (B2) irregular designs.

3.2.2. SMOOTHNESS CLASSES OF FUNCTIONS. Let β, L be positive numbers, and let $k=\lfloor\beta\rfloor$ be the maximal integer such that $k<\beta$. Consider the smoothness classes of functions $\Sigma(\beta,L)$ and $\Sigma_N(\beta,L)$ defined in Sections 1.2 and 2.7.6 respectively. For $0<h<1/2$ define the following classes of uniformly bounded smooth functions $g:[0,1]\to[0,1]$:

$$\Sigma(\beta,L,h)=\Sigma(\beta,L)\cap\left\{g(x_1):\ h\le g(x_1)\le 1-h,\ 0\le x_1\le 1\right\},$$

and the classes of uniformly bounded smooth functions $g:[0,1]^N\to[0,1]$:

$$\Sigma_N(\beta,L,h)=\Sigma_N(\beta,L)\cap\left\{g(x_1):\ h\le g(x)\le 1-h,\ x\in[0,1]^N\right\}.$$

3.2.3. ASSUMPTIONS ON THE DOMAINS. The domain G is a compact in K having a smooth boundary. We study separately the following two cases.

A boundary fragment. Denote by $\mathcal{G}_0=\mathcal{G}_0(\gamma,L,h)$ the class of all domains G satisfying

$$G=\left\{x=(x_1,x_2)\in K:\ 0\le x_1\le 1,\ 0\le x_2\le g(x_1)\right\}$$

where $g\in\Sigma(\gamma,L,h)$, $\gamma>0$.

If $G\in\mathcal{G}_0$ then the "roof" of the boundary $\Gamma=\partial G$ is defined by a

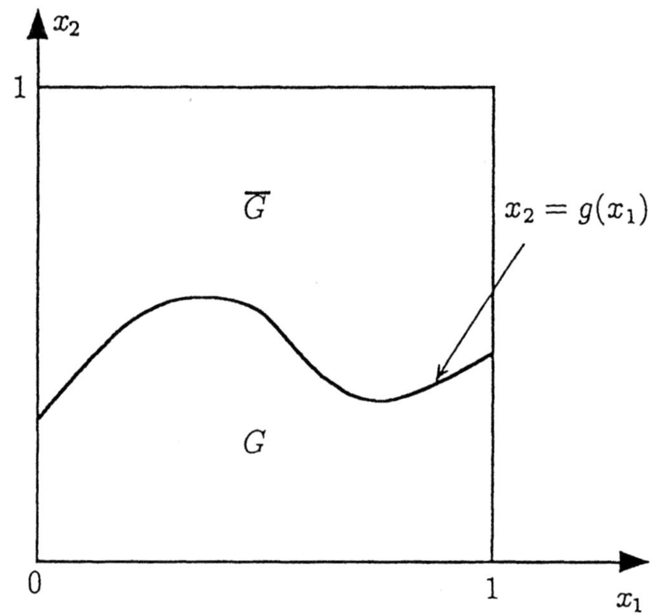

Fig. 3.1. A boundary fragment.

bona fide function

$$g: [0,1] \to [0,1].$$

Images with such domains G are called boundary fragments. They can be obtained by cutting small squares out of the original image and further rescaling. The small squares can be associated with some special coordinate systems depending on the local boundary structure. A method for extraction of boundary fragments from the general image is proposed in Section 5.3. The model of boundary fragment is the same as "upper-lower regions" model of Rudemo, Skovgaard and Stryhn (1990), and the model considered by Carlstein and Krishnamoorthy (1992).

General case. The general classes $\mathcal{G}$ of domains are defined as sets of compacts G in K bounded by smooth curves. As above, we associate the smoothness of domains in $\mathcal{G}$ with a parameter $\gamma \geq 0$: $\mathcal{G}=\mathcal{G}(\gamma)$. Several definitions of general classes $\mathcal{G}$ are presented in Chapter 5.

3.2.4. CLASSES OF IMAGES. Let γ, a, b be positive numbers, and $0 \leq \beta \leq \infty$. Define the class of grey-scale images $\Phi_{0,\gamma}$ as the set of all functions f on K that can be represented as

$$f(x) = f_0(x) I\{x \in \bar{G}\} + f_1(x) I\{x \in G\}$$

where $G \in \mathcal{G}=\mathcal{G}(\gamma)$, and

$$\sup_{x \in \bar{G}} f_0(x) \leq a < b \leq \inf_{x \in G} f_1(x), \qquad 0 < a < b < 1. \tag{3.2}$$

For $0<\beta<\infty$ define $\Phi_{\beta,\gamma}$ as the subset of $\Phi_{0,\gamma}$ where $f_0, f_1 \in \Sigma_2(\beta, L_1)$, $0<L_1<\infty$.

Classes of binary images are denoted by $\Phi_{\infty,\gamma}$. They consist of the indicator functions

$$f(x) = I\{x \in G\}$$

with $G \in \mathcal{G}=\mathcal{G}(\gamma)$.

The classes of boundary fragments $\Phi^0_{\beta,\gamma}$, $0 \leq \beta \leq \infty$, are defined in the same way as $\Phi_{\beta,\gamma}$ with condition $G \in \mathcal{G}_0(\gamma, L, h)$ instead of $G \in \mathcal{G}$. The classes of grey-scale boundary fragments are those classes $\Phi^0_{\beta,\gamma}$ for which $\beta<\infty$, while $\Phi^0_{\infty,\gamma}$ is the class of <u>binary boundary fragments</u>.

Thus, in the notation $\Phi_{\beta,\gamma}$ the first subscript is responsible for the image smoothness while the second one is for smoothness of the domain's boundary. The superscript 0 in $\Phi^0_{\beta,\gamma}$ is used only to indicate the case of boundary fragment.

3.2.5. RISK FUNCTIONS FOR EDGE ESTIMATION. Let $\hat{G}_n$ be an estimator of the domain G based on observations (3.1). This means that $\hat{G}_n$ is a closed subset of K measurable with respect to observations (X_i,Y_i). We measure the error of edge estimation by some distance $d(G,\hat{G}_n)$ between G and $\hat{G}_n$. In this book we consider the following two distances.

<u>Hausdorff distance</u>. Define the Hausdorff distance between the two closed compact sets G_1 and G_2 as

$$d_\infty(G_1,G_2) = \max\left\{\max_{x\in G_1} \rho(x,G_2),\ \max_{x\in G_2} \rho(x,G_1)\right\}$$

where $\rho(x,G)$ is the Euclidean distance between a point x and a closed set G.

Note that for $G_1,G_2\in\mathcal{G}_0$ the Hausdorff distance is equivalent to the L_∞-distance between the corresponding functions $g_1(x_1)$ and $g_2(x_1)$. Keeping this in mind, we use the following notation in the case of boundary fragments:

$$d_\infty(G_1,G_2) = \sup_{0\le x_1\le 1} |g_1(x_1)-g_2(x_1)|.$$

<u>Measure of symmetric difference</u>. Let $G_1\Delta G_2$ be the symmetric difference between compact sets G_1 and G_2. Define the distance as the Lebesgue measure of $G_1\Delta G_2$,

$$d_1(G_1,G_2) = \mathrm{mes}(G_1\Delta G_2).$$

For $G_1,G_2\in\mathcal{G}_0$ this equals to the L_1-distance between the corresponding functions $g_1(x_1)$ and $g_2(x_1)$:

$$d_1(G_1,G_2) = \int_0^1 |g_1(x_1)-g_2(x_1)|dx_1.$$

For binary images $f(x)=I\{x\in G\}$ denote by P_G the probability measure generated by observations (X_i,Y_i), $i=1,\dots,n$, satisfying (3.1) where G is the image domain. Denote by E_G the expectation with respect to P_G.

In this book we deal with the risk functions for edge estimation defined by

$$R(\hat{G}_n,G) = E_G\left(w(\psi_n^{-1}d(G,\hat{G}_n))\right)$$

(cf. Section 1.7) where the distances are $d=d_1$, $d=d_\infty$ and the loss functions are $w(u)=u^2$ (squared loss), $w(u)=|u|$ (absolute value loss), $w(u)=I\{|u|\ge C\}$, $C>0$ (indicator loss); ψ_n is a positive normalizing sequence.

3.2.6. RISK FUNCTIONS FOR IMAGE ESTIMATION. Let $T_n(x)$ be an image estimator, i.e. a function on K measurable with respect to observations (3.1). The error of T_n is determined in a standard way by means of a nonnegative risk function $R(T_n,f)$. We use the L_2-risk with squared loss function as defined in Example 1.7.1:

$$R(T_n,f) = \psi_n^{-2}\mathrm{MISE}(T_n,f) = \psi_n^{-2}E_f\left(\int_K (f(x)-T_n(x))^2dx \right) .$$

Here T_n is an arbitrary estimate of f and E_f is the expectation with respect to the distribution of and observations (X_i,Y_i), $i=1,\dots,n$, satisfying (3.1).

3.2.7. MINIMAX RATE OF CONVERGENCE FOR EDGE ESTIMATION. Let us adapt the definitions given in Section 2.1.2 to edge estimation. In this problem the estimated parameter ϑ is the set G, and the pseudometrics are $d=d_1$ and $d=d_\infty$. Hence, we say that the positive sequence ψ_n is <u>I-minimax rate of convergence</u> for edge estimation if there exist two positive numbers $p_0<1$ and C such that

(3.3)
$$\liminf_{n\to\infty} \inf_{\hat{G}_n} \sup_{G\in\mathcal{G}} P_G(\psi_n^{-1} d(G,\hat{G}_n) \ge C) \ge p_0$$

($\inf_{\hat{G}_n}$ is the infimum over all edge estimators), and there exists a sequence of estimators G_n^* such that

(3.4)
$$\lim_{C\to\infty} \limsup_{n\to\infty} \sup_{G\in\mathcal{G}} P_G(\psi_n^{-1} d(G,G_n^*) \ge C) = 0.$$

Here d is a distance between compact sets, e.g. the Hausdorff distance or the measure of symmetric difference. The estimator G_n^* satisfying (3.4) is called <u>I-optimal edge estimator</u>.

Similarly, ψ_n is the <u>S-minimax rate of convergence</u> for edge estimation if for any n large enough

(3.5)
$$C_0 \le \inf_{\hat{G}_n} \sup_{G\in\mathcal{G}} E_G(\psi_n^{-2} d^2(G,\hat{G}_n)) \le C_1$$

with some positive C_0, C_1, and G_n^* is the optimal estimator if for all n

(3.6)
$$\sup_{G\in\mathcal{G}} E_G(\psi_n^{-2} d^2(G,G_n^*)) \le C_1 .$$

For the case of boundary fragment definitions are analogous (replace $\mathcal{G}$ by $\mathcal{G}_0$ everywhere). Also the <u>A-minimax rate of convergence</u> can be defined if one substitutes $\psi_n^{-1} d(G,G_n^*)$ instead of $\psi_n^{-2} d^2(G,G_n^*)$ under the expectation sign in (3.5) and (3.6).

3.2.8. MINIMAX RATE OF CONVERGENCE FOR IMAGE ESTIMATION. For image estimation problem we use only the squared loss function and the L_2-norm. Therefore, following the notation of Section 2.1.2 we have to define S-minimax rates and S-optimal estimators. For brevity we omit the prefix "S-" since it does not lead to misunderstanding. A positive sequence ψ_n is called <u>minimax rate of convergence</u> for the class of images Φ if for any n large enough

$$C_0 \le \inf_{T_n} \sup_{f\in\Phi} \psi_n^{-2} \mathrm{MISE}(T_n, f) \le C_1 \tag{3.7}$$

with some positive C_0, C_1. Here Φ is a class of images, e.g. one of the classes described in Section 3.2.4 and $\inf_{T_n}$ is the infimum over all image estimators. An estimator $\hat{f}_n$ is called <u>optimal image estimator</u> if for any n

$$\sup_{f\in\Phi} \psi_n^{-2} \mathrm{MISE}(\hat{f}_n, f) \le C_1 \tag{3.8}$$

where ψ_n is the minimax rate.

3.3. LOWER BOUNDS ON THE ACCURACY OF ESTIMATES

To find minimax rates of convergence and optimal image and edge estimators we have to prove two sorts of results. First, the lower bounds, i.e. (3.3) or the left inequalities in (3.5) and (3.7). Second, the upper bounds (3.4), (3.6) or (3.8) for particular estimators. In this section we discuss the lower bounds. Since it turns out that the rates of convergence strongly depend on the choice of design we consider separately the lower bounds under regular and irregular designs (see the definitions in Section 3.2.1).

Here we prove I-lower bounds for the case of boundary fragments, i.e. we get inequalities of the form

$$\liminf_{n\to\infty} \inf_{\hat{G}_n} \sup_{G\in\mathcal{G}_0} P_G(\psi_n^{-1} d(G,\hat{G}_n) \ge C) \ge p_0. \tag{3.9}$$

One can easily verify that this entails I-lower bounds with the same rate ψ_n for the general classes $\mathcal{G}$ of (two-dimensional) domains considered in Chapter 5. In view of Proposition 2.6.1 the S-lower bounds also follow.

<u>Binary image, regular design</u>. Lower bounds under the regular designs (A1) and (A2) are rather simple since the least possible estimation error turns out to be proportional to the step size

$1/\sqrt{n}$ of the regular grid.

3.3.1. THEOREM. Let the design satisfy (A1) or (A2). Then the lower bound (3.9) holds for the distances $d=d_1$ and $d=d_\infty$ with the rate $\psi_n=1/\sqrt{n}$.

Proof. Consider the domains

$$G=\{(x_1,x_2):\ x_2\le c\} \text{ and } G_1'=\{(x_1,x_2):\ x_2\le c+1/(2\sqrt{n})\}$$

where c is a constant, 0<c<1. This constant can be chosen so that the slice $G\backslash G'$ contains no design points under the design (A1) and contains no design points with probability at least 1/2 under the design (A2). Thus, Assumption 2.2.1 holds with $\lambda=0$ and $p=1/2$. Application of Proposition 2.2.2 shows that with probability at least 1/4 it is impossible to distinguish between the hypotheses P_G and $P_{G'}$ using the observations (3.1). This entails (3.9) with $\psi_n=1/\sqrt{n}$.

Theorem 3.3.1 does not answer the question whether the rate $1/\sqrt{n}$ is achievable under the designs (A1) and (A2). The answer is positive (see Chapter 4 for further details).

Binary image, irregular design. Now we proceed to the minimax lower bounds under the designs (B1) and (B2). To present the result in the most transparent form assume that the noise ξ_i in (3.1) is Gaussian.

3.3.2.THEOREM. Let the errors ξ_i be zero-mean Gaussian random variables with variance $\sigma^2>0$ and let the design satisfy (B1) or (B2). Then (3.9) holds with rates

(i) $\psi_n=(1/n)^{\gamma/(\gamma+1)}$ if d is the measure of symmetric difference d_1;

(ii) $\psi_n=(\log n/n)^{\gamma/(\gamma+1)}$ if d is the Hausdorff distance d_∞.

3.3.3.REMARK. The lower bound (3.9) implies

$$\liminf_{n\to\infty}\ \inf_{\hat{G}_n}\ \sup_{G\in\mathcal{G}_0} E_G(\psi_n^{-q}d^q(G,\hat{G}_n)) \ge C_0,\quad q>0. \tag{3.10}$$

(cf. Proposition 2.6.1).

We shall give the extension of Theorem 3.3.2 to non-Gaussian case later in this section.

3.3.4. PROOF OF THEOREM 3.3.2. Denote

$$\delta_n = \begin{cases} (1/n)^{1/(\gamma+1)} & \text{if } d=d_1, \\ (\log n/n)^{1/(\gamma+1)} & \text{if } d=d_\infty. \end{cases}$$

In this proof we assume without loss of generality that $m=m_n=1/\delta_n$ is an integer. Let $u_\ell=\ell\delta_n$, $\ell=0,1,\dots,m$. Consider the intervals

$$U_1=[0,u_1),\ U_2=[u_1,u_2),\ \dots,\ U_m=[u_{m-1},1].$$

Denote by b_ℓ the center of the interval U_ℓ. Let ϕ be the basic function for the class $\Sigma(\gamma,L)$ defined in Section 2.4.2. Then the functions

$$g_{\ell n}(x_1) = 1/2 + \delta_n^\gamma \phi(\delta_n^{-1}(x_1-b_\ell)),\ \ell=1,\dots,m,$$

belong to the class $\Sigma(\gamma,L)$ by Proposition 2.4.3. These functions have the same form as the functions (2.53), up to the additive constant 1/2. We add 1/2 here to ensure that for n large $g_{\ell n}$ satisfies the condition

$$h \le g_{\ell n}(x_1) \le 1-h$$

which is required in the definition of $\mathcal{G}_0=\mathcal{G}_0(\gamma,L,h)$.

Denote

$$c = \int \phi(x_1)dx_1.$$

Introduce the domains

$$G_0 = [0,1] \times [0,1/2],$$

and

$$G_\ell = \{(x_1,x_2)\in K:\ 0 \le x_2 \le g_{\ell n}(x_1)\},\ \ell=1,\dots,m.$$

Note that G_ℓ coincides with G_0 everywhere except the slice

$$A_\ell = \left\{(x_1,x_2):\ u_{\ell-1}\le x_1<u_\ell\right\},\quad u_\ell = \ell\delta_n .$$

Also,

$$\text{mes}(G_\ell \backslash G_0) = \int g_{\ell n}(x_1)dx_1 - \frac{1}{2} = c\delta_n^{\gamma+1}.$$

Assume first that $d=d_\infty$. Set $M=m$,

$$\vartheta_\ell = g_{\ell n}(\cdot), \ \vartheta_0 \equiv 0,$$

$$P_{\vartheta_\ell} = P_{G_\ell}, \quad \ell=0,1,\ldots,M.$$

We apply Theorem 2.5.3 and (2.30). Let us check that Assumption 2.5.1 holds for the proposed choice of $\vartheta_0, \vartheta_1, \ldots, \vartheta_M$.

Clearly, Assumption 2.5.1 (a) is satisfied with

$$s_n = \frac{1}{2}\phi_* \delta_n^\gamma = \frac{1}{2}\phi_* (\log n/n)^{\gamma/(\gamma+1)}$$

where $\phi_* = \max_x \phi(x)$.

It remains to prove 2.5.1 (b). Since ξ_i are Gaussian the likelihood ratios are of the form

$$\Lambda(\vartheta_0, \vartheta_\ell; X^{(n)}) = dP_{G_0}/dP_{G_\ell} =$$

$$= \exp\left(- \sum_{i=1}^n I\{X_i \in G_\ell \backslash G_0\} Y_i + \text{card}(i: X_i \in G_\ell \backslash G_0)/2 \right).$$

It is obvious that under the designs (B1), (B2) the random numbers of design points $\Lambda_\ell = \text{card}(i: X_i \in G_\ell \backslash G_0)$ have expectations

$$E(\Lambda_\ell) = n \ \text{mes}(G_\ell \backslash G_0) = cn\delta_n^{\gamma+1}.$$

Hence, under the hypothesis ϑ_ℓ the likelihood ratio can be represented as

$$\Lambda(\vartheta_0, \vartheta_\ell; X^{(n)}) = \exp(\Delta_\ell - cn\delta_n^{\gamma+1}/2)$$

where

$$\Delta_\ell = - \sum_{i=1}^n I\{X_i \in G_\ell \backslash G_0\} \xi_i + (E(\Lambda_\ell) - \Lambda_\ell)/2$$

are zero mean random variables.

For $d=d_\infty$ we have

$$n\delta_n^{\gamma+1} = \log n.$$

Hence (2.32) is satisfied if we choose ϕ_* (respectively, c) small enough.

Let us prove (2.33). We consider only the design (B1) (the modification for (B2) is simple). Under (B1) Λ_ℓ can be represented as the sum of i.i.d. Bernouillian random variables:

$$\Lambda_\ell = \sum_{i=1}^{n} \tilde{\eta}_{\ell n}(X_i)$$

where $\tilde{\eta}_{\ell n}(X_i) = I\{X_i \in G_\ell \backslash G_0\}$,

$$P(\tilde{\eta}_{\ell n}(X_i) = 1) = c\delta_n^{\gamma+1}, \quad \forall\ i,\ell.$$

Hence the distribution of Λ_ℓ does not depend on ℓ. By the central limit theorem

(3.11) $$P(E(\Lambda_\ell) - \Lambda_\ell \geq 0) = P(E(\Lambda_1) - \Lambda_1 \geq 0) \to 1/2, \quad \text{as } n \to \infty.$$

Now we act as in Example 2.7.2. We have

(3.12) $$P_{G_\ell}(\Delta_\ell \geq 0) \geq$$

$$\geq P\left(-\sum_{i=1}^{n} I\{X_i \in G_\ell \backslash G_0\}\xi_i \geq 0, \quad E(\Lambda_\ell) - \Lambda_\ell \geq 0 \right) =$$

$$= E\left(P\left(-\sum_{i=1}^{n} I\{X_i \in G_\ell \backslash G_0\}\xi_i \geq 0 \mid X_1,\dots,X_n\right) I\{E(\Lambda_\ell) - \Lambda_\ell \geq 0\} \right).$$

For fixed $X_1,\dots,X_n$ the random variables $-\sum_{i=1}^{n} I\{X_i \in G_\ell \backslash G_0\}\xi_i$ are zero-mean Gaussian, possibly degenerate. Hence

(3.13) $$P\left(-\sum_{i=1}^{n} I\{X_i \in G_\ell \backslash G_0\}\xi_i \geq 0 \mid X_1,\dots,X_n\right) = 1/2 \quad \text{(a.s.)},$$

and in the degenerate case this probability is 1 a.s. It follows from (3.11) - (3.13) that uniformly in ℓ

$$P_{G_\ell}(\Delta_\ell \geq 0) \geq 1/8$$

for n large. This proves (2.33).

Now assume that $d=d_1$. We apply again Theorem 2.5.3, although with another definition of hypotheses ϑ_j. Set $M=2^m$,

$$\vartheta_j = g(\cdot,\bar{\omega}_j) = f(\cdot,\bar{\omega}_j) + \frac{1}{2}, \; j=0,1,\ldots,M,$$

where $f(\cdot,\bar{\omega}_j)$ is the function defined in (2.61), and $\bar{\omega}_0,\bar{\omega}_1,\ldots,\bar{\omega}_M$ are the binary vectors from Lemma 2.7.4. Denote

$$G_j = \{(x_1,x_2)\in K\colon \; 0 \leq x_2 \leq g(x_1,\bar{\omega}_j)\}, \; j=0,1,\ldots,M,$$

$$P_{\vartheta_j}=P_{G_j}, \; j=0,1,\ldots,M.$$

Let us check Assumption 2.5.1. First, the inequalities (2.31) are satisfied since

$$d(\vartheta_j,\vartheta_k) = \int |f(x_1,\bar{\omega}_j)-f(x_1,\bar{\omega}_k)|dx_1 =$$

$$= c\delta_n^{\gamma+1}\rho_H(\bar{\omega}_j,\bar{\omega}_k) \geq \frac{cm}{16n} = 2s_n$$

where $s_n=(c/32)n^{-\gamma/(\gamma+1)}$. The likelihood ratio is

$$\Lambda(\vartheta_0,\vartheta_j;X^{(n)}) = \exp(\Delta_j - n\,\mathrm{mes}(G_j\backslash G_0)/2).$$

Here

$$n\,\mathrm{mes}(G_j\backslash G_0) \leq cmn\delta_n^{\gamma+1} \leq \lambda \log M$$

with some $0<\lambda<1$ if c is small enough. Hence, (2.32) in Assumption 2.5.1 is satisfied. Finally, to check (2.33) we use the same argument similar to (3.11)-(3.13). Hence, Theorem 2.5.3 applies, and the I-lower bound in d_1-metric holds with $\psi_n=n^{-\gamma/(\gamma+1)}$. □

Non-Gaussian noise and arbitrary design. Theorem 3.2.1 can be generalized in several ways. First, it remains valid for non-Gaussian noise distributions under very mild assumptions.

Second, the minimax lower bounds for edge estimation in d_1-distance hold under arbitrary design.

Consider the observations (3.1) with binary image f and i.i.d. errors ξ_i. Some assumptions on the distribution function F of ξ_i are necessary to get nontrivial lower bounds. We assume throughout this section that the supports of F(y) and F(y-1) have a non-empty intersection (otherwise it is possible to distinguish between the domain and the background without error).

Denote by

$$H^2(t) = \int ((dF(u+t))^{1/2} - (dF(u))^{1/2})^2$$

the squared Hellinger distance between the distributions of ξ_i and $t+\xi_i$, $t\in R^1$.

3.3.5. ASSUMPTION. $H^2(t) < 2$, $\forall t\in R^1$.

3.3.6. THEOREM. Let Assumption 3.3.5 be satisfied, and let the design points $(X_1,\ldots,X_n)$ be random variables independent of $(\xi_1,\ldots,\xi_n)$. Then the lower bound (3.10), q=1, holds for the distance $d=d_1$ with $\psi_n=(1/n)^{\gamma/(\gamma+1)}$.

This theorem is the special case of the more general Theorem 5.1.1 which is proved in Chapter 5. The extension to higher dimensions (N>2) will also be given in Theorem 5.1.1.

Grey-scale image. Now we extend the lower bounds to the case of grey-scale images. Consider the classes of images described in Section 3.2.4. As in edge estimation it is sufficient to prove lower bounds for grey-scale boundary fragments, i.e. to get the inequalities of the form

$$\liminf_{n\to\infty} \inf_{T_n} \sup_{f\in\Phi^0_{\beta,\gamma}} \psi_n^{-2} \mathrm{MISE}(T_n,f) \ge C_0. \tag{3.14}$$

3.3.7. THEOREM. Let the image f in (3.1) belong to $\Phi^0_{\beta,\gamma}$, and let $(X_1,\ldots,X_n)$ be an arbitrary design independent of $(\xi_1,\ldots,\xi_n)$. If the distribution of noises ξ_i satisfies

$$\limsup_{t\to 0} H^2(t)/t^2 < +\infty , \tag{3.15}$$

then the lower bound (3.14) holds, with

$$\psi_n = \max\left\{(1/n)^{\gamma/(2(\gamma+1))},\ (1/n)^{\beta/(2(\beta+1))}\right\},\ \gamma\ge 1,\ 0<\beta<\infty .$$

The constant C_0 in (3.14) does not depend on T_n and on the design.

Proof. It suffices to prove the inequalities

$$\liminf_{n\to\infty} \inf_{T_n} \sup_{f\in\Phi^0_{\beta,\gamma}} n^{\gamma/(\gamma+1)} \mathrm{MISE}(T_n,f) \ge C_0 \tag{3.16}$$

and

$$\liminf_{n\to\infty} \inf_{T_n} \sup_{f\in\Phi^0_{\beta,\gamma}} n^{\beta/(\beta+1)} \mathrm{MISE}(T_n,f) \ge C_0 \tag{3.17}$$

for some constant $C_0>0$ which does not depend on the design. Note that (3.16) follows from Theorem 3.3.6. In fact,

$$\inf_{T_n} \sup_{f\in\Phi^0_{\beta,\gamma}} n^{\gamma/(\gamma+1)} \mathrm{MISE}(T_n,f) \ge$$

$$\ge \inf_{T_n} \sup_{f\in\Phi^0_{\infty,\gamma}} n^{\gamma/(\gamma+1)} \mathrm{MISE}(T_n,f) \ge \inf_{\hat G_n} \sup_{G\in\mathcal{G}_0} n^{\gamma/(\gamma+1)} E_G(d_1(G,\hat G_n))$$

since for every estimate T_n

$$\mathrm{MISE}(T_n,f) = E_f\left(\int (I\{x:\ x_2\le g(x_1)\} - T_n(x))^2 dx\right) \ge$$

$$\ge (1/2)\ E_f\left(\int (I\{x:\ x_2\le g(x_1)\} - I\{x\in\hat G_n\})^2 dx\right) = (1/2)\ E_G(d_1(G,\hat G_n))$$

where $\hat G_n = \{x:\ T_n(x) > 1/2\}$.

Let us prove (3.17). By definition of the class of grey-scale boundary fragments $\Phi^0_{\beta,\gamma}$ the functions f that belong to this class have no edges in the rectangles $[0,1]\mathrm{x}[0,h]$ and $[0,1]\mathrm{x}[1-h,1]$. Hence there exists a square $K_1\subseteq K$ with the sides of length $h/2$ such that for every $f\in\Phi^0_{\beta,\gamma}$ the restriction

$f_{K_1}(x)=f(x)I\{x\in K_1\}$ of f to K_1 is smooth on K_1 and belongs to the restriction of the class $\Sigma_2(\beta,L)$ to K_1. This entails

$$\inf_{T_n} \sup_{f\in\Phi^0_{\beta,\gamma}} n^{\beta/(\beta+1)} \mathrm{MISE}(T_n, f) \geq$$

$$\geq \inf_{T_n} \sup_{f\in\Phi^0_{\beta,\gamma}} n^{\beta/(\beta+1)} E_f\left(\int_{K_1} (f_{K_1}(x) - T_n(x))^2 dx\right) \geq$$

$$\geq \inf_{T_n} \sup_{f\in\Sigma_2(\beta,L)} n^{\beta/(\beta+1)} E_f\left(\int_{K_1} (f_{K_1}(x) - T_n(x))^2 dx\right).$$

Using the techniques of Assouad's lemma one can easily prove that the last expression is bounded from below by some constant that does not depend on the design (see Section 2.7.6 and Remark 2.8.5). □

CHAPTER 4. OPTIMAL IMAGE AND EDGE ESTIMATION FOR BOUNDARY FRAGMENTS

Here we define optimal image and edge estimators that attain the lower bounds of Chapter 3. For edge estimation we use the modified version of PPE, while for image estimation the two-dimensional LPE with appropriately chosen random windows is proposed. To concentrate on the main ideas we consider only the case of boundary fragments in Gaussian noise. Generalizations and extensions are delayed to Chapter 5.

4.1. OPTIMAL EDGE ESTIMATION

Let us construct optimal edge estimators for boundary fragments of the class $\mathcal{G}_0(\gamma, L, h)$. This will be done by combining piecewise-polynomial approximations with maximum likelihood estimation. The optimal estimator will be defined similarly for distances $d=d_1$ and $d=d_\infty$ (only the values of bin width δ_n will differ).

Let

$$\delta_n = \begin{cases} (1/n)^{1/(\gamma+1)} & \text{if } d=d_1, \\ (\log n/n)^{1/(\gamma+1)} & \text{if } d=d_\infty. \end{cases}$$

Denote $M=M_n=1/\delta_n$ and assume without loss of generality that M is an integer. Set, as usually, $u_\ell=\ell\delta_n$, $\ell=0,1,\dots,M$,

$$U_1=[0,u_1),\ U_2=[u_1,u_2),\dots,\ U_M=[u_{M-1},1],$$

$A_\ell=U_\ell \times R^1$ (Note that each slice A_ℓ is infinite along the axis x_2). Introduce the domains

$$D_\ell = G \cap A_\ell\ ,\quad \ell=1,\dots,M.$$

In each slice A_ℓ define the parametric family of domains

$$B_\ell(\theta) = \left\{(x_1,x_2)\in A_\ell:\ 0 \le x_2 \le \theta_0+\theta_1(x_1-u_{\ell-1})+\dots+\theta_k(x_1-u_{\ell-1})^k\right\}$$

where $k=\lfloor\gamma\rfloor$ is the maximal integer such that $k<\gamma$, and $\theta=(\theta_0,\dots,\theta_k)$ is a vector of real parameters. Hence, for some values of θ the curve

$$x_2=\theta_0+\theta_1(x_1-u_{\ell-1})+\ldots+\theta_k(x_1-u_{\ell-1})^k, \quad u_{\ell-1}\le x_1<u_\ell,$$

may leave the unit square K. Let

$$\Theta^{(K)} = \left\{ \theta: \quad 0 \le \theta_0+\theta_1(x_1-u_{\ell-1})+\ldots+\theta_k(x_1-u_{\ell-1})^k \le 1, \ \forall x_1\in U_\ell \right\}.$$

Note that $\Theta^{(K)}$ does not depend on ℓ. For any $\theta\in\Theta^{(K)}$ the domain $B_\ell(\theta)$ is correctly defined and located within the square K.

Consider the maximum likelihood estimator of θ based on observations in a slice A_ℓ. To avoid technical problems we introduce the maximum likelihood estimator on the discrete set

$$\Theta_n = \left\{ \theta = (m_0\delta_n^{\gamma},\ m_1\delta_n^{\gamma-1}, \ldots,\ m_k\delta_n^{\gamma-k}) \right\} \cap \Theta^{(K)}$$

where $m=(m_0,m_1,\ldots,m_k)$ is a $(k+1)$tuple of integers.

For Gaussian distribution of noise the maximum likelihood estimator $\hat\theta_\ell=(\hat\theta_{0\ell},\ldots,\hat\theta_{k\ell})$ for the slice A_ℓ is a solution of

$$J_\ell(\hat\theta_\ell) = \max_{\theta\in\Theta_n} J_\ell(\theta)$$

where

$$J_\ell(\theta) = \sum_i I\{X_i\in B_\ell(\theta)\}(Y_i - \tfrac{1}{2}).$$

Define the edge estimator

$$\hat g(x_1) = \hat g_\ell(x_1) = \hat\theta_{0\ell}+\hat\theta_{1\ell}(x_1-u_{\ell-1})+\ldots+\hat\theta_{k\ell}(x_1-u_{\ell-1})^k,$$

$$u_{\ell-1} \le x_1 < u_\ell, \quad \ell=1,\ldots,M.$$

Note that this is the piecewise-polynomial estimator of type (1.13) (put $\tilde\theta_{n\ell}=(\hat\theta_{0\ell},\delta_n\hat\theta_{1\ell},\ldots,\delta_n^k k!\,\hat\theta_{k\ell})$ in (1.13) to get the expression for $\hat g_\ell$).

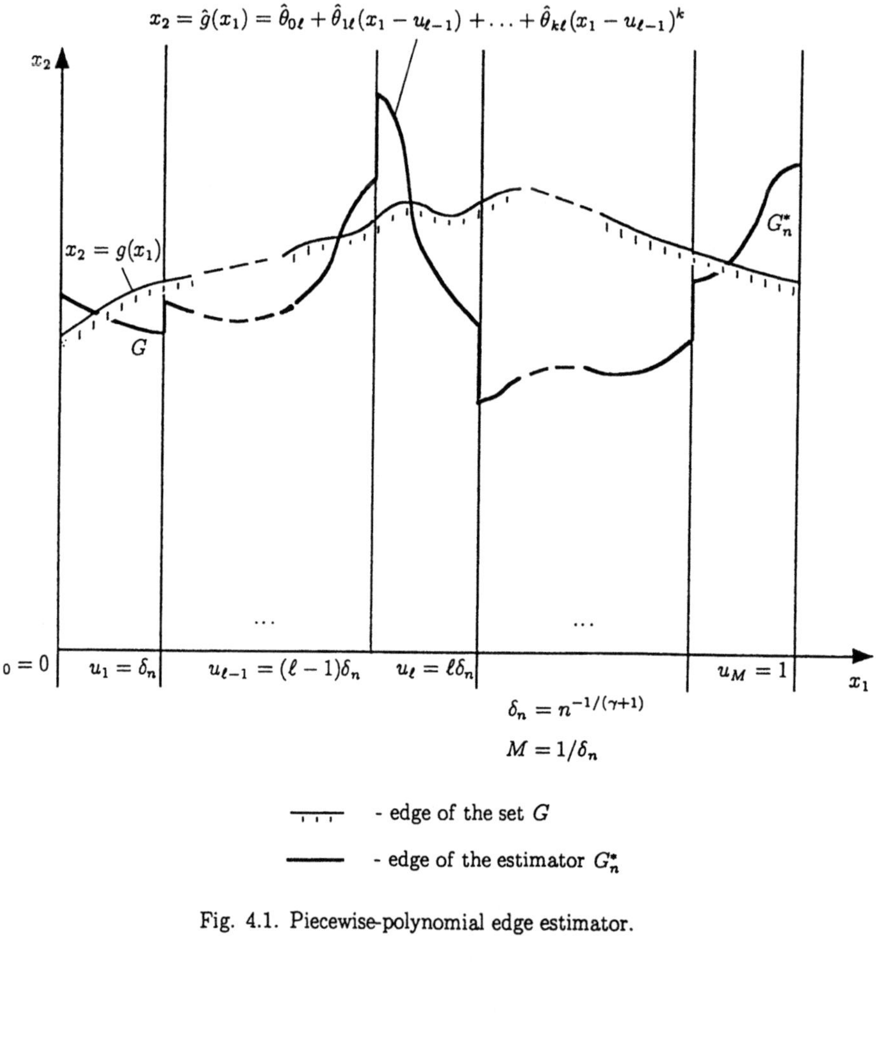

Fig. 4.1. Piecewise-polynomial edge estimator.

Finally, denote by G_n^* the closure of the set

$$\left\{(x_1,x_2):\ 0 \le x_2 \le \hat{g}(x_1),\ 0 \le x_1 \le 1\right\}.$$

The estimator G_n^* appears to be the optimal edge estimator. The following theorem shows that for the case of irregular designs (B1), (B2) the estimator G_n^* satisfies (3.4), and thus the I-lower rates of convergence in Theorem 3.3.2 are in fact the I-minimax rates.

4.1.1. THEOREM. Let ξ_i be i.i.d. zero mean Gaussian random variables and let the design satisfy (B1) or (B2) (see Section 3.2.1). Then for the estimator G_n^* the equality (3.4) holds where $\mathcal{G}=\mathcal{G}_0(\gamma,L,h)$, $\gamma\ge 1$, and

(i) $\psi_n=(1/n)^{\gamma/(\gamma+1)}$ if d is the measure of symmetric difference d_1;

(ii) $\psi_n=(\log n/n)^{\gamma/(\gamma+1)}$ if d is the Hausdorff distance d_∞.

4.2. PRELIMINARY LEMMAS

Each vector of parameters $\theta=(\theta_0,\dots,\theta_k)\in\Theta_n$ is characterized by a set of integers $m=(m_0,\dots,m_k)$, this correspondence being denoted as $\theta=\theta(m)$. With this notation the maximum likelihood estimator is $\hat{\theta}_\ell=\theta(\hat{m}_\ell)$.

Introduce the norm

$$\|m\| = \max_{0\le j\le k} |m_j|.$$

Define the Taylor approximation $B_\ell^{(0)}$ of the set $D_\ell=G\cap A_\ell$:

$$B_\ell^{(0)} = \left\{(x_1,x_2)\in A_\ell:\ 0 \le x_2 \le g_\ell^{(0)}(x_1)\right\}$$

where

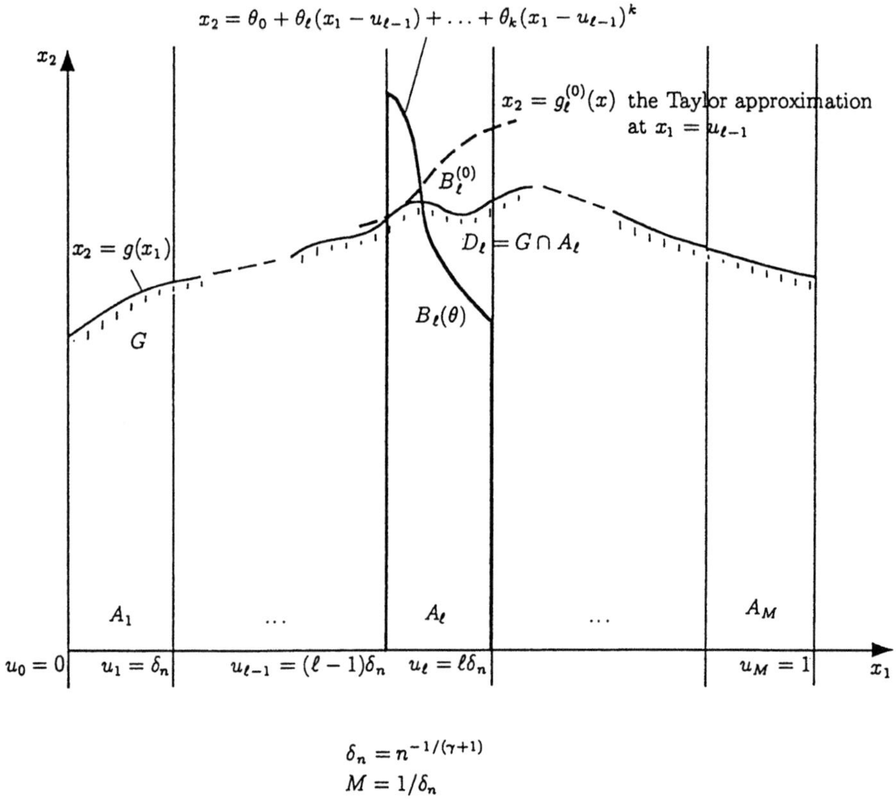

Fig. 4.2. The domains used in the proof of Theorem 4.1.1.

$$g_\ell^{(0)}(x_1) = \theta_0^{(0)}+\theta_1^{(0)}(x_1-u_{\ell-1})+\dots+\theta_k^{(0)}(x_1-u_{\ell-1})^k,$$

$$(\theta_0^{(0)}, \theta_1^{(0)},\dots, \theta_k^{(0)}) = (m^{(0,0)}\delta_n^\gamma, m^{(0,1)}\delta_n^{\gamma-1},\dots, m^{(0,k)}\delta_n^{\gamma-k})$$

and

$$m^{(0,j)} = \left[\frac{g^{(j)}(u_{\ell-1})}{j!\ \delta_n^{\gamma-j}}\right], \quad j=0,\dots,k.$$

Let $m_\ell^{(0)}=(m^{(0,0)}, m^{(0,1)},\dots, m^{(0,k)})$.

4.2.1. LEMMA. There exist such positive r_i, $i=1,\dots,5$, that uniformly in $\ell=1,\dots,M$,

$$(4.1)\qquad r_1\delta_n^{\gamma+1}\|m-m_\ell^{(0)}\|\le d_1(B_\ell(\theta(m)),B_\ell^{(0)})\le r_2\delta_n^{\gamma+1}\|m-m_\ell^{(0)}\|,$$

$$(4.2)\qquad \delta_n^{\gamma+1}(r_3\|m-m_\ell^{(0)}\|-r_4)\le d_1(B_\ell(\theta(m)),D_\ell)\le\delta_n^{\gamma+1}(r_5\|m-m_\ell^{(0)}\|+r_4).$$

Proof. To prove (4.1) note that

$$d_1(B_\ell(\theta(m)),B_\ell^{(0)}) = \delta_n^\gamma\int_{u_{\ell-1}}^{u_\ell}\Big|\sum_{j=0}^{k}(m_j - m^{(0,j)})\ ((x_1-u_{\ell-1})/\delta_n)^j\Big|dx_1=$$

$$= \delta_n^{\gamma+1}\int_0^1\Big|\sum_{j=0}^{k}(m_j - m^{(0,j)})y^j\Big|dy.$$

Now, (4.1) is proved since the last integral is bounded from above and from below by $r_i\|m-m_\ell^{(0)}\|$, $i=1,2$. Inequalities (4.2) follow from (4.1) and from the definition of $\mathcal{G}_0$ in view of the triangle inequality. □

4.2.2. LEMMA. Under the conditions of Theorem 4.1.1 the log-likelihood ratios are of the form

$$(4.3)\qquad Z_\ell(\theta) = Z_\ell\Big(\theta, (X_1, Y_1), \ldots, (X_n, Y_n)\Big) =$$

$$= \log\Big(dP_{B_\ell(\theta)}/dP_{D_\ell}\Big) = \Delta_\ell(\theta) - nd_1(B_\ell(\theta), D_\ell)/2$$

where the random variables Δ_ℓ are independent for different ℓ, $E_G(\Delta_\ell)=0$ and for all z with $|z|$ small enough

$$(4.4)\qquad E_G(\exp(z\Delta_\ell)) \le \exp(\varkappa_0 z^2 nd_1(B_\ell(\theta), D_\ell))$$

with a positive $\varkappa_0$ which does not depend on n, ℓ, θ and G.

Proof. The log-likelihood ratio can be written explicitly:

$$(4.5)\qquad Z_\ell(\theta) = \sum_i I\{X_i \in B_\ell(\theta)\backslash D_\ell\}\xi_i - \sum_i I\{X_i \in D_\ell\backslash B_\ell(\theta)\}\xi_i - \Lambda_\ell/2$$

where Λ_ℓ = card $\{i: X_i \in B_\ell(\theta)\Delta D_\ell\}$ with the mean value $\bar{\Lambda}_\ell = E_G(\Lambda_\ell) = nd_1(B_\ell(\theta), D_\ell)$. It remains to verify the inequality (4.4). As the variables ξ_i in (4.5) are Gaussian and independent of the design, we have

$$(4.6)\qquad E_G(\exp(z\Delta_\ell)) = E_G\Big(\exp\{\ \Lambda_\ell z^2/2 - (\Lambda_\ell - \bar{\Lambda}_\ell)z/2\}\Big) =$$

$$= \exp\{\bar{\Lambda}_\ell z^2/2\}\ E_G\Big(\exp\{(\Lambda_\ell - \bar{\Lambda}_\ell)(z^2 - z)/2\}\Big).$$

Under the design (B1) Λ_ℓ is the sum of independent Bernouillian random variables with variance which is less than $\bar{\Lambda}_\ell$. This implies

$$(4.7)\qquad E_G\Big(\exp\{(\Lambda_\ell - \bar{\Lambda}_\ell)z_0\}\Big) \le \exp(z_0^2\bar{\Lambda}_\ell)$$

for $|z_0| \le 1$. Under the design (B2) the same is true if M is a multiple of $n^{1/2}$. If M is not a multiple of $n^{1/2}$ then under (B2) there is a dependence between Λ_ℓ and $\Lambda_{\ell-1}$ but (4.7) still holds (we don't give the details for this case here).

Finally, (4.4) follows from (4.6) and (4.7). □

4.3. PROOF OF THEOREM 4.1.1

Below we use $c_1, c_2, \ldots,$ to denote positive constants independent of n.

(i) $d=d_1$. For any $R>0$ we have

$$P_G\left(\|\hat{m}_\ell - m_\ell^{(0)}\| > R\right) \le P_G\left(\max_{\|m-m_\ell^{(0)}\|>R} \left(Z_\ell(\theta(m)) - Z_\ell(\theta(m_\ell^{(0)}))\right) > 0\right). \tag{4.8}$$

Introduce the spherical layers of vectors with integer coordinates:

$$L(r) = \{m: Rr < \|m-m_\ell^{(0)}\| \le R(r+1)\} \cap \{m: \theta(m) \in \Theta^{(K)}\}$$

where $r=1,2,\ldots$. Clearly, $L(r)$ contains less than $c_1 r^{k} R^{k+1}$ points.

Substituting the representation (4.3) for the log-likelihood ratio into the right-hand side of (4.8) and using inequalities (4.1), (4.2), we get

$$P_G\left(\max_{m\in L(r)} \left(Z_\ell(\theta(m)) - Z_\ell(\theta(m_\ell^{(0)}))\right) > 0\right) \le \tag{4.9}$$

$$\le P_G\left(\max_{m\in L(r)} \left(\Delta_\ell(\theta(m)) - \Delta_\ell(\theta(m_\ell^{(0)}))\right) > (\varkappa_3 Rr - 2\varkappa_4)/2\right) \le$$

$$\le \sum_{m\in L(r)} P_G\left(\Delta_\ell(\theta(m)) - \Delta_\ell(\theta(m_\ell^{(0)})) > (\varkappa_3 Rr - 2\varkappa_4)/2\right).$$

For any $m\in L(r)$ the exponential Chebyshev's inequality applies

$$P_G\left(\Delta_\ell(\theta(m)) - \Delta_\ell(\theta(m_\ell^{(0)})) > (\varkappa_3 Rr - 2\varkappa_4)/2\right) \le$$

$$\le \exp(-z(\varkappa_3 Rr - 2\varkappa_4)/2)\; E_G\left[\exp\left(z(\Delta_\ell(\theta(m)) - \Delta_\ell(\theta(m_\ell^{(0)})))\right)\right] \le$$

$$\le \exp(-\tfrac{z}{2}(r_3Rr-2r_4))\exp\Big(r_0z^2n[d_1(B_\ell(\theta(m)),D_\ell)+d_1(B_\ell(\theta(m_\ell^{(0)})),D_\ell)]\Big)$$

where the Hölder inequality and (4.4) have been used. If we substitute this into the right-hand side of (4.9) and sum up over $m\in L(r)$ then

$$P_G\Big(\max_{m\in L(r)}\Big[Z_\ell(\theta(m)) - Z_\ell(\theta(m_\ell^{(0)}))\Big] > 0\Big) \le$$

$$\le c_1r^kR^{k+1}\exp(r_0z^2(r_5R(r+1)+2r_4)-z(r_3Rr-2r_4)/2)$$

where Lemma 4.2.1 has been used again. Finally, we take $z=r_3/(4r_5r_0)$ and obtain

$$P_G\Big(\max_{m\in L(r)}\Big[Z_\ell(\theta(m))-Z_\ell(\theta(m_\ell^{(0)}))\Big] > 0\Big) \le \tag{4.10}$$

$$\le c_2(rR)^k\exp(-r_3^2rR/(16r_5r_0)).$$

Summing up (4.10) in $r>R$ we find that the right-hand side of (4.8) does not exceed $c_3\exp(-R/c_4)$. Again by (4.2)

$$P_G(\delta_n^{-\gamma}d_1(G,G_n^*)\ge C) \le P_G\Big(\delta_n\sum_{\ell=1}^{M}(r_5\|\hat m_\ell-m_\ell^{(0)}\|+r_4) > C\Big) =$$

$$= P_G\Big(M^{-1}\sum_{\ell=1}^{M}\|\hat m_\ell-m_\ell^{(0)}\| > (C-r_4)/r_5\Big).$$

To complete the proof of (i) note that under the design (B1) the random variables $\|\hat m_\ell-m_\ell^{(0)}\|$ are independent and in view of (4.8),(4.10) their distributions have exponentially decreasing tails. Hence, the law of large numbers applies to $M^{-1}\sum_{\ell=1}^{M}\|\hat m_\ell-m_\ell^{(0)}\|$, and the last probability tends to zero as $C\to\infty$. A similar argument, with suitable modifications, applies for the design (B2) (see the remarks after (4.7)).

(ii) $d=d_\infty$. Calculations similar to (4.8)-(4.10) lead to

$$(4.11)\qquad P_G(\|\hat{m}_\ell - m_\ell^{(0)}\| > R) \le n^{-R/c_5}$$

for R large enough. In fact, since $\delta_n = (\log n/n)^{1/(\gamma+1)}$ the inequalities which are analogous to (4.8) and (4.10) now take the form

$$P_G(\|\hat{m}_\ell - m_\ell^{(0)}\| > R) \le \sum_{r\ge 1} P_G\left(\max_{m\in L(r)} \left(Z_\ell(\theta(m)) - Z_\ell(\theta(m_\ell^{(0)}))\right) > 0\right) \le$$

$$\le \sum_{r\ge 1} c_6 (rR)^{k} \exp(-c_7 rR \log n) \le n^{-R/c_5}.$$

The rest of proof is based on the definition of G_n^*. We have

$$(4.12)\qquad P_G(\delta_n^{-\gamma} d_\infty(G, G_n^*) \ge C) =$$

$$= P_G\left(\delta_n^{-\gamma} \max_{1\le\ell\le M} \sup_{u_{\ell-1}\le x_1 < u_\ell} |\hat{g}(x_1) - g(x_1)| \ge C\right) \le$$

$$\le \sum_{\ell=1}^{M} P_G\left(\delta_n^{-\gamma} \sup_{u_{\ell-1}\le x_1 < u_\ell} (|\hat{g}(x_1) - g_\ell^{(0)}(x_1)| + |g(x_1) - g_\ell^{(0)}(x_1)|) \ge C \right).$$

Since $g\in\Sigma(\gamma, L)$ we get

$$(4.13)\qquad \sup_{u_{\ell-1}\le x_1 < u_\ell} |g(x_1) - g_\ell^{(0)}(x_1)| \le \delta_n^{\gamma} L/k! + \delta_n^{\gamma}(k+1)$$

where the last summand appears due to the discreteness of the coefficients of $g_\ell^{(0)}$. By definition of $\hat{g}$

$$(4.14)\qquad \sup_{u_{\ell-1}\le x_1 < u_\ell} |\hat{g}(x_1) - g_\ell^{(0)}(x_1)| \le \delta_n^{\gamma}(k+1)\|\hat{m}_\ell - m_\ell^{(0)}\|.$$

Inequalities (4.11)-(4.14) show that the last sum in (4.12) does not exceed

$$\sum_{\ell=1}^{M} P_G(\|\hat{m}_\ell - m_\ell^{(0)}\| \geq (C-L/k!)/(k+1)-1) \leq n^{-C/c_8}$$

if C is large enough. Hence C may be chosen in such a way that, e.g.

$$P_G(\delta_n^{-\gamma} d_\infty(G,G_n^*) \geq C) \leq n^{-2}. \tag{4.15}$$

The theorem is proved. □

4.3.1. REMARK. The reason for assuming that $\gamma \geq 1$ is quite obvious with the design (B2): otherwise the bin width δ_n is less than the step $1/\sqrt{n}$ of the regular grid and the observations in different slices A_ℓ are no longer independent. For (B1) this hurdle does not exist and the theorem holds with $0<\gamma<1$ as well.

4.3.2. REMARK. Since the distribution of maximum likelihood estimator $\hat{m}_\ell$ has exponentially decreasing tails it is straightforward to get inequalities of type (3.6) for squared loss (or any power loss) with the same rates of convergence.

4.3.3. REMARK. Theorem 4.1.1 remains valid under the zebra design which is a modification of (B2) defined as follows. Let

$$K^{odd} = \bigcup_{\ell \text{ odd}} A_\ell, \qquad K^{even} = \bigcup_{\ell \text{ even}} A_\ell,$$

$$G^{odd} = G \cap K^{odd}, \quad G^{even} = G \cap K^{even}.$$

Then $(B2^{odd})$ and $(B2^{even})$ are the restrictions of design (B2) on K^{odd} and K^{even} respectively. "Zebra" design will be used below as an auxiliary tool.

With zebra design the estimator $\hat{g}$ should be properly defined on the "empty" slices. Therefore we assume that under $(B2^{odd})$ the estimator is of the form

$$\hat{g}^{odd}(x_1) = \hat{g}_\ell(x_1), \quad u_{\ell-1} \leq x_1 < u_{\ell+1}, \quad \ell \text{ odd},$$

and the similar definition is given under $(B2^{even})$.

4.4. OPTIMAL IMAGE ESTIMATION

Consider the model (3.1) where $f \in \Phi^{0}_{0,\gamma}$ is a grey-scale boundary fragment and ξ_i are zero mean Gaussian random variables.

We propose a two-step procedure of image estimation. On the first step we find some estimators G_n^* and $\bar{G}_n^*$ of the domain and background respectively (in other words this is the edge estimation step). The second step consists in smoothing of image values separately on the sets G_n^* and $\bar{G}_n^*$.

For technical reasons suppose that n is even and that the sample is split into two equal parts. The first subsample is used to construct the edge estimator. Smoothing is performed from n/2 observations of the second subsample.

Some remarks are in order here since the way of splitting the sample will depend on the design. We consider now only the designs (B1) and (B2). For (B1) the splitting of sample does not change the type of design: it is again of type (B1) on each subsample. We assume that under (B1) the first and second subsamples contain observations with odd and even numbers i respectively. For (B2) the designs on subsamples are no longer of type (B2). Define the splitting rule for (B2) in such a way that the first and second subsamples have zebra designs ($B2^{odd}$) and ($B2^{even}$) respectively.

To define our edge estimator for grey-scale images we should know a threshold parameter, i.e. a constant between a and b. In the following we suppose without loss of generality that such constant is given by (a+b)/2. It will be clear that for our purposes the threshold can be replaced by its consistent estimate converging fast enough. One easily verifies that such estimates exist in many cases, and we will not discuss this in detail here.

4.4.1. EDGE ESTIMATION STEP. Define the new observations

$$Y_i' = I\{Y_i > (a+b)/2\}. \tag{4.16}$$

This may be written as

(4.17) $$Y_i' = I\{X_i \in G\}\tilde{\zeta}_i + I\{X_i \in \bar{G}\}\tilde{\eta}_i$$

where $\tilde{\zeta}_i, \tilde{\eta}_i$ are independent, although, nonidentically distributed Bernoullian random variables, e.g.

$$P_f\left(\tilde{\zeta}_i = 1\right) = P_f\left(f(X_i) + \xi_i > (a+b)/2 \,|\, X_i \in G\right) \geq$$

$$\geq P\left(b - (a+b)/2 > -\xi_i\right) > 1/2.$$

Thus, the original grey-scale image with additive noise is transformed to a binary image with multiplicative nonidentically distributed noise. This suggests to apply the edge estimation approach as proposed in Section 4.1 for binary images. However, it is impossible to calculate the likelihood function related to observations (4.17) since the distributions of $\tilde{\zeta}_i, \tilde{\eta}_i$ depend on f. Therefore, we will try the same estimator as in Section 4.1 based on the new observations. This will be no longer the maximum likelihood estimator in a slice. We show now that such an estimator converges with the optimal rate, i.e. the following analogue of Theorem 4.1.1 holds.

4.4.2. THEOREM. Let ξ_i be i.i.d. zero mean Gaussian random variables and let the design satisfy (B1) or (B2). Let G_n^* be the estimator of Section 4.1 based on new observations (X_i, Y_i') from the first subsample. Then uniformly over the grey-scale boundary fragments $f \in \Phi^0_{0,\gamma}$, $\gamma \geq 1$,

(4.18) $$\limsup_{n\to\infty} \sup_{f \in \Phi^0_{0,\gamma}} E_f(\psi_n^{-1} d(G, G_n^*)) \leq C_1$$

for $d = d_1$ and $d = d_\infty$ with ψ_n as in Theorem 4.1.1. The constant C_1 does not depend on f.

Note that no smoothness conditions on image values are imposed here. The image can be arbitrarily bad provided $G \in \mathcal{G}_0 = \mathcal{G}_0(\gamma, L, h)$ and (3.2) holds.

Proof follows the same lines as in Sections 4.2 and 4.3 with $Z_\ell(\theta) = Z_\ell(\theta, \{(X_i, Y_i')\})$ instead of $Z_\ell(\theta, (X_1, Y_1), \ldots, (X_n, Y_n))$ where the indices "i" in (X_i, Y_i') correspond to the first subsample. The only point in the proof to be modified is Lemma 4.2.2 which is replaced by the next result.

4.4.3. LEMMA. Under the conditions of Theorem 4.4.2 there exists a positive constant c such that

$$-cnd_1(B_\ell(\theta), D_\ell) \geq Z_\ell(\theta) - \Delta_\ell(\theta) \geq -2cnd_1(B_\ell(\theta), D_\ell)$$

where the random variables Δ_ℓ are independent for different ℓ, $E_f(\Delta_\ell)=0$ and for all z small enough

$$E_f(\exp(z\Delta_\ell)) \leq \exp(\kappa_0 z^2 nd_1(B_\ell(\theta), D_\ell)) \tag{4.19}$$

with a positive κ_0 which does not depend on n, ℓ, θ and f.

Proof. Denote $B_\ell = B_\ell(\theta)$, $Z_\ell = Z_\ell(\theta)$. We have

$$Z_\ell = \sum_i \Big(I\{X_i \in B_\ell \cap \bar{D}_\ell\} - I\{X_i \in D_\ell \cap \bar{B}_\ell\} \Big) (Y_i' - 1/2).$$

Hence

$$Z_\ell = \Delta_{\ell 1} + \Delta_{\ell 2} + E(Z_\ell) \tag{4.20}$$

where

$$\Delta_{\ell 1} = Z_\ell - E(Z_\ell | X_1, \ldots, X_n) =$$

$$= \sum_i \Big(I\{X_i \in B_\ell \cap \bar{D}_\ell\} - I\{X_i \in D_\ell \cap \bar{B}_\ell\} \Big) (Y_i' - E(Y_i' | X_i)),$$

$$\Delta_{\ell 2} = E(Z_\ell | X_1, \ldots, X_n) - E(Z_\ell).$$

It is easy to see that $\Delta_{\ell 1}$, $\Delta_{\ell 2}$ are sums of independent bounded random variables with $E(\Delta_{\ell 1}) = E(\Delta_{\ell 2}) = 0$,

$$E(\Delta_{\ell 1}^2 | X_1, \ldots, X_n) \leq 2 \operatorname{card}\{i : X_i \in B_\ell \Delta D_\ell\},$$

$$E(\Delta_{\ell 1}^2) \le 2nd_1(B_\ell, D_\ell), \quad E(\Delta_{\ell 2}^2) \le nd_1(B_\ell, D_\ell)$$

since $E(\text{card}\{i: X_i \in B_\ell \Delta D_\ell\}) = nd_1(B_\ell, D_\ell)$ under (B1) and (B2). This entails (4.19) with $\Delta_\ell = \Delta_{\ell 1} + \Delta_{\ell 2}$ by application of standard techniques related to Chebyshev's inequality. Now,

$$E(Y_i' | X_i) = P(\xi_i > (a+b)/2 - f(X_i) \,|X_i).$$

For $X_i \in G$

$$1/2+\pi_1 = P\Big(\xi_i > -(b-a)/2\Big) \le E(Y_i'|X_i) \le P\Big(\xi_i > (a+b)/2 - 1\Big) = 1/2+\pi_1',$$

and for $X_i \in \overline{G}$

$$1/2-\pi_2 = P\Big(\xi_i > (a+b)/2\Big) \le E(Y_i'|X_i) \le P\Big(\xi_i > (b-a)/2\Big) = 1/2-\pi_2'$$

where $0 < \pi_1, \pi_1', \pi_2, \pi_2' < 1/2$. Thus,

$$E(Y_i'|X_i) - 1/2 = \pi(X_i)\Big[I\{X_i \in G\} - I\{X_i \in \overline{G}\}\Big]$$

where $\min(\pi_1, \pi_2') \le \pi(X_i) \le \max(\pi_2, \pi_1')$. Hence,

$$E(Z_\ell | X_1, \ldots, X_n) =$$

$$= \sum_i \Big(I\{X_i \in B_\ell \cap \overline{D}_\ell\} - I\{X_i \in D_\ell \cap \overline{B}_\ell\}\Big) \; (E(Y_i'|X_i) - 1/2) =$$

$$= -\sum_i \pi(X_i) I\{X_i \in B_\ell \Delta D_\ell\},$$

and

$$-\frac{n}{2}\min(\pi_1, \pi_2') d_1(B_\ell, D_\ell) \ge E(Z_\ell) \ge -\frac{n}{2}\max(\pi_2, \pi_1') d_1(B_\ell, D_\ell).$$

This together with (4.20) entails the lemma. □

We have stated Theorem 4.4.2 for the absolute value loss function (cf. Remark 4.3.2). Using (4.19) one can prove that the theorem holds for any power loss function as well.

4.4.4. SMOOTHING STEP. In the following it will be convenient to compare edge estimation with painting into black and white. The points $x \in G_n^*$ and $x \in \bar{G}_n^*$ are called white and black respectively. We say that two points x, x' have the same colour if either $x, x' \in G_n^*$ or $x, x' \in \bar{G}_n^*$.

The idea of this step is to use the two-dimensional version of LPE with appropriately chosen random windows. The windows depend on the results of edge estimation.

Set $h_n = n^{-1/(2\beta+2)}$, $0<\beta<\infty$. For fixed $x \in K$ define

$$m_n(x) = \text{card } \mathcal{I}_n(x)$$

where $\mathcal{I}_n(x) = \left\{ i : X_i \in \mathcal{X}^2,\ |X_i - x| \le h_n,\ X_i \text{ has the same colour as } x \right\}$, $\mathcal{X}^2$ denotes the second subsample of design points. The data $\{X_i, i \in \mathcal{I}_n(x)\}$ are in a random window of diameter $\le 2h_n$ around x.

Let $\ell = \lfloor \beta \rfloor$ be the maximal integer such that $\ell < \beta$, and let $\mathcal{P}_\ell$ be the set of polynomials of degree ℓ on K. Denote by $p_{n,\ell}$ the least squares polynomial of degree ℓ for regression with the data $\{(X_i, Y_i),\ i \in \mathcal{I}_n(x)\}$, i.e. the solution of the following minimization problem

$$\min_{p \in \mathcal{P}_\ell} \sum_{i \in \mathcal{I}_n(x)} (Y_i - p(X_i))^2$$

If $0<\beta\le 1$, then $\ell=0$ and $p_{n,\ell}$ is the running mean in our random window:

$$p_{n,0}(x) = \begin{cases} m_n(x)^{-1} \sum_{i \in \mathcal{I}_n(x)} Y_i & \text{if } m_n(x) \ne 0, \\ 0 & \text{otherwise.} \end{cases}$$

Define the image estimator

$$(4.21) \qquad \hat{f}_n(x) = p_{n,\Delta}(x)\ I\{0 \le p_{n,\Delta}(x) \le 1\}.$$

4.4.5. THEOREM. Let the assumptions of Theorem 4.4.2 be satisfied. Assume that for the edge estimation step $\delta_n = n^{-1/(\gamma+1)}$. Then for image estimator (4.21) the upper bound (3.8) holds with $\Phi = \Phi^0_{\beta,\gamma}$, and

$$\psi_n = \max\left\{(1/n)^{\gamma/(2(\gamma+1))},\ (1/n)^{\beta/(2(\beta+1))}\right\},\ \gamma \ge 1,\ 0 < \beta < \infty.$$

$$\psi_n = (1/n)^{\gamma/(2(\gamma+1))},\ \gamma \ge 1,\ \beta = \infty.$$

4.5. PROOF OF THEOREM 4.4.5

The statement of this theorem for $\beta = \infty$ follows directly from Theorem 4.1.1 (i). Thus assume that β is finite. The proof will be given for the simplest case of $\Delta = 0$ (respectively, $0 < \beta \le 1$). Extension to $\Delta > 0$ follows the same lines, although it requires more technicalities of usual kind that we would like to avoid here. Readers who are interested in a detailed development for $\Delta > 0$ may combine the main stages of proof described here with the proof of Theorem 1.6.6.

For $x \in K$ define the random events

$$\Xi_n(x) = \{m_n(x) \ge c_\Xi n h_n^2\}$$

where c_Ξ is a constant, and

$$\Omega_{in}(x) = \{X_i \in \mathcal{X}^2 \cap \mathcal{U}_n(x)\}$$

where $\mathcal{U}_n(x) = \{z: |z - x| \le h_n\} \cap (G \Delta G_n^*)\}$.

4.5.1. LEMMA. Under the conditions of Theorem 4.4.5 uniformly in $x \in K$

$$E\left[(\mathrm{mes}\,\mathcal{U}_n(x))^j\right] = O\left(n^{-j\gamma/(\gamma+1)} h_n^j\right),\ n \to \infty,\ j = 1, 2.$$

<u>Proof</u>. Consider the design (B1) since (B2) is treated similarly. Let $x=(x_1,x_2)\in K$ be fixed. Define

$$\mathcal{L} = \left\{ \ell : [u_{\ell-1}, u_\ell] \cap [x_1-h_n, x_1+h_n] \neq \emptyset \right\}.$$

Then $\operatorname{card}(\mathcal{L}) \le M_n h_n + 1$ where $M_n=\delta_n^{-1}$, $\delta_n = n^{-1/(\gamma+1)}$. Applying (4.2) one gets

$$\delta_n^{-\gamma} h_n \text{ mes } \mathcal{U}_n(x) \le \delta_n^{-\gamma} h_n^{-1} \int_{x_1-h_n}^{x_1+h_n} |\hat{g}(t)-g(t)|\, dt \le$$

$$\le \delta_n^{-\gamma} h_n^{-1} \sum_{\ell\in\mathcal{L}} \int_{u_{\ell-1}}^{u_\ell} |\hat{g}(t)-g(t)|dt \le (M_n h_n)^{-1} \sum_{\ell\in\mathcal{L}} (r_5 \| \hat{m}_\ell - m_\ell^{(0)} \| + r_4).$$

But for $\| \hat{m}_\ell - m_\ell^{(0)} \|$ the exponential inequality (4.9) holds (with P_f instead of P_G, see Lemma 4.4.3 and remarks before it). Since the random variables $\| \hat{m}_\ell - m_\ell^{(0)} \|$ are independent for different ℓ this proves the lemma. Moreover, the assertion of the lemma holds for any $j>0$. □

4.5.2. LEMMA. Under the conditions of Theorem 4.4.5 there exists a constant c_Ξ such that uniformly in $x\in K$

$$P\left(\bar{\Xi}_n(x)\right) = o(h_n^2), \quad n \to \infty.$$

<u>Proof</u>. Let $x\in G$ (the case of $x\in\bar{G}$ is considered similarly). We have

$$E(m_n(x)|W^1) = (n/2) \text{ mes}(z\in G_n^*:\ |z-x| \le h_n) \ge$$

$$\ge (n/2)\Big(\text{mes}(z\in G:\ |z-x| \le h_n) - \text{mes } \mathcal{U}_n(x)\Big)$$

where $E(\cdot|W^1)$ denotes the conditional expectation for the fixed first subsample. Since the edge function g is Lipschitz there exists a positive constant c_G such that

$$\text{mes}(z\in G:\ |z-x| \le h_n) \ge c_G h_n^2$$

for $n \ge n_0$ where n_0 depends only on L, β, h. Thus,

$$E(m_n(x)|W^1) \ge c_G n h_n^2/2 - (n/2)\ \text{mes}\ \mathcal{U}_n(x).$$

Set $c_\Xi = c_G/4$. Then

$$P(\overline{\Xi}_n(x)) \le P\left(m_n(x) - E(m_n(x)|W^1) < -c_G n h_n^2/4 - (n/2)\ \text{mes}\ \mathcal{U}_n(x)\right) \le$$

$$\le P\left(m_n(x) - E(m_n(x)|W^1) < -c_G n h_n^2/8\right) +$$

$$+ P\left((n/2)\ \text{mes}\ \mathcal{U}_n(x) > c_G n h_n^2/8\right) \le$$

$$\le E\left[\ P\left(m_n(x) - E(m_n(x)|W^1) < -c_G n h_n^2/8\ |\ W^1\right)\ \right] + O\left(n^{-2\gamma/(\gamma+1)} h_n^{-2}\right)$$

by Lemma 4.5.1. To complete the proof it remains to note that m_n is the sum of n bounded random variables that are conditionally independent for fixed W^1, and the conditional variance of each summand is less than πh_n^2. This together with the Bernstein's inequality implies that

$$P\left(m_n(x) - E(m_n(x)|W^1) < -c_G n h_n^2/8\ |\ W^1\right) \le 2\exp(-c_9 n h_n^2)$$

with a constant c_9 that does not depend on W^1. □

Remind that $0 \le f, \hat{f}_n \le 1$. It follows from Lemma 4.5.2 that

$$E_f\left(\int I\{\overline{\Xi}_n(x)\}(f(x)-\hat{f}_n(x))^2 dx\right) = o(h_n^2).$$

Hence,

(4.22) $\quad \text{MISE}(\hat{f}_n, f) \le$

$$\le E_f(d_1(G,G_n^*)) + E_f\Big(\int_{G\cap\bar{G}_n^*} I\{\Xi_n(x)\}(f(x)-\hat{f}_n(x))^2dx\Big)+$$

$$+ E_f\Big(\int_{G\cap G_n^*} I\{\Xi_n(x)\}(f(x)-\hat{f}_n(x))^2dx\Big) + o(h_n^2).$$

Let us evaluate the last integral in (4.22). Define

$$\mathcal{I}_{n1}(x) = \{i\in\mathcal{I}_n(x):\ X_i\in G\} \quad\text{and}\quad \mathcal{I}_{n0}(x) = \{i\in\mathcal{I}_n(x):\ X_i\in\bar{G}\cap G_n^*\}.$$

Then $\hat{f}_n(x)=f_{n0}(x)+f_{n1}(x)$ where on Ξ_n

$$f_{nj}(x) = m_n(x)^{-1}\sum_{i\in\mathcal{I}_{nj}} Y_i,\ j=0,1.$$

Note that on Ξ_n

$$(4.23)\qquad (f_{n0}(x) - f(x))^2 \le \frac{2}{c_\Xi^2 n^2 h_n^4}\Big[\Big(\sum_{i\in\mathcal{I}_{n0}} \xi_i\Big)^2 + (\text{card }\mathcal{I}_{n0})^2\Big].$$

By Lemma 4.5.1

$$(4.24)\qquad E_f\Big(\Big(\sum_{i\in\mathcal{I}_{n0}}\xi_i\Big)^2\Big) \le E_f\Big(E\Big[\Big(\sum_i \xi_i I\{\Omega_{in}\}\Big)^2|\mathcal{W}^1\Big]\Big) \le$$

$$\le c_{10}E(\text{mes }\mathcal{U}_n(x)) = O(h_n^2)$$

and

$$(4.25)\qquad E((\text{card }\mathcal{I}_{n0})^2) \le E\Big(E\Big[\Big(\sum_i I\{\Omega_{in}\}\Big)^2|\mathcal{W}^1\Big]\Big) =$$

$$= E\Big(E\Big[\Big(\sum_i (I\{\Omega_{in}\}-E(I\{\Omega_{in}\}|\mathcal{W}^1))\Big)^2|\mathcal{W}^1\Big] + E\Big(\sum_i E(I\{\Omega_{in}\}|\mathcal{W}^1)\Big)^2\Big) \le$$

$$\le c_{11}(E(n \text{ mes } \mathcal{U}_n(x)) + E(n^2 \text{mes}^2 \mathcal{U}_n(x))) = o(n^{-\beta/(\beta+1)})$$

if $\gamma \ge 1$. From (4.23) - (4.25) we obtain

$$E_f\left(\int_{G \cap G^*} I\{\Xi_n(x)\}(f(x)-f_{n0}(x))^2 dx\right) = o(n^{-\beta/(\beta+1)}). \tag{4.26}$$

Next, on Ξ_n

$$(f_{n1}(x) - f(x))^2 \le \frac{2}{c_{\Xi}^2 n^2 h_n^4} \left[\left(\sum_{i \in \mathcal{J}_{n1}} \xi_i\right)^2 + \left(\sum_{i \in \mathcal{J}_{n1}} (f_1(X_i) - f_1(x))\right)^2\right].$$

Hence, by Hölder condition on f_1

$$E_f\left((f_{n1}(x) - f(x))^2 I(x \in G \cap G_n^*) \right) \le \tag{4.27}$$

$$\le \frac{c_{12}}{n^2 h_n^4}\left[E\left((\sum_i \xi_i I\{X_i \in G \cap \mathcal{X}^2 : |X_i - x| \le h_n\})^2\right) + \right.$$

$$\left. + L^2 h_n^{2\beta} E\left((\sum_i I\{X_i \in G \cap \mathcal{X}^2 : |X_i - x| \le h_n\})^2\right)\right] =$$

$$= O\left(\frac{1}{n^2 h_n^4} (n\mu_n + h_n^{2\beta}(n\mu_n + n^2\mu_n^2))\right) = O(n^{-\beta/(\beta+1)})$$

where

$$\mu_n = \begin{cases} \text{mes}\{z \in G : \ |z-x| \le h_n\} & \text{under (B1)} \\ \text{mes}\{z \in G^{\text{even}} : |z-x| \le h_n\} & \text{under (B2)}. \end{cases}$$

In both cases $\mu_n = O(h_n^2)$. It follows from (4.26) and (4.27) that

$$E_f\left(\int_{G \cap G^*} I\{\Xi_n(x)\}(f(x) - \hat{f}_n(x))^2 dx\right) = O(n^{-\beta/(\beta+1)}).$$

The same is true for the integral over $\overline{G} \cap \overline{G}_n^*$. This together with (4.22) and (4.18) for $d=d_1$ proves the theorem. □

CHAPTER 5. GENERALIZATIONS AND EXTENSIONS

5.1. HIGH-DIMENSIONAL BOUNDARY FRAGMENTS. NON-GAUSSIAN NOISE

The results of Chapters 3 and 4 can be generalized for the images defined on N-dimensional cube $K=[0,1]^N$, $N\geq 2$. Also, assumptions on the noise distribution can be relaxed. To consider these points we need some more notation.

First, extend the definition of $\mathcal{G}_0$ to higher dimensions. Let $\gamma\geq 1$, $L>0$, $0<h<1/2$ be fixed. Denote by $\mathcal{G}_{0,N}$ the class of all domains G in $K=[0,1]^N$ satisfying

$$G = \left\{ (x_1,\ldots,x_N)\in K:\ 0\leq x_N\leq g(x_1,\ldots,x_{N-1}) \right\}$$

where $g\in\Sigma_{N-1}(\gamma,L,h)$.

Define the classes of N-dimensional boundary fragments $\Phi^0_{\beta,\gamma,N}$, $0\leq\beta\leq\infty$, similarly to $\Phi^0_{\beta,\gamma}$ (see Section 3.2.4), with the only difference that $\mathcal{G}_0$, $\Sigma_2(\beta,L_1)$ are replaced by $\mathcal{G}_{0,N}$, $\Sigma_N(\beta,L_1)$.

Along with the N-dimensional extension of (3.1) we consider the multiplicative noise model for binary boundary fragments. For multiplicative model it will be more convenient to assume that the values of f are -1 and 1 on the background and on the domain respectively, i.e.

$$f(X_i) = I\{X_i\in G\} - I\{X_i\in \bar{G}\}.$$

With this definition of f the multiplicative model is of the form

$$(5.1)\qquad Y_i = \xi_i f(X_i),\quad i=1,\ldots,n,$$

where

$$\xi_i = \begin{cases} 1 \text{ w.p. } \bar{p}_0, \\ -1 \text{ w.p. } 1-\bar{p}_0, \end{cases}$$

$$1/2 < \bar{p}_0 < 1$$

(cf. (1.47)).

The model (5.1) is related to image transmission through a binary symmetric communication channel. With probability p_0 the image values are transmitted without error, and with probability

$1-p_0$ they are inverted.

Lower bounds for estimation of N-dimensional boundary fragments are given by the following extensions of Theorems 3.3.6 and 3.3.7.

5.1.1. THEOREM. Let the design $(X_1,\ldots X_n)$, $X_i\in[0,1]^N$, $i=1,\ldots,n$, be arbitrary but independent of $(\xi_1,\ldots,\xi_n)$. Then for the model (3.1) with $f\in\Phi^0_{\infty,\gamma,N}$ under Assumption 3.3.5 and for the model (5.1)

$$\liminf_{n\to\infty}\ \inf_{\hat{G}_n}\ \sup_{G\in\mathcal{G}_{0,N}} E_G(\psi_n^{-q}d_1^q(G,\hat{G}_n)) \ge C_0 > 0,\ \forall q\ge 1,$$

where $\psi_n=n^{-\gamma/(\gamma+N-1)}$.

<u>Proof</u> is based on the argument similar to Assouad's lemma (Theorem 2.6.4). It suffices to prove the theorem for q=1 since

$$E_G(\psi_n^{-q}d_1^q(G,\hat{G}_n)) \ge \Big(E_G(\psi_n^{-1}d_1(G,\hat{G}_n))\Big)^q,\ q>1.$$

Let $\phi(t)$ be infinitely many times differentiable function on R^{N-1} with support $(-1/2,1/2)^{N-1}$ such that $0<\phi(t)\le 1$ (ϕ is the (N-1)-dimensional analogue to the basic function defined in Section 2.4.2).

Introduce a partition of $[0,1]^{N-1}$ into the disjoint cubes Q_j, $j=1,\ldots,M$, $M=m^{N-1}$, with edges of length $1/m$ where m is an integer. Let b_j be the center of Q_j.

Set

$$m=[n^{1/(\gamma+N-1)}].$$

Consider the family of edge functions

$$g^r(t,\bar{\omega}) = r/m^\gamma + (\bar{c}/m^\gamma)\sum_{j=1}^{M}\omega_j\phi(m(t-b_j))$$

where $t=(x_1,\ldots,x_{N-1})$, $\bar{\omega} = (\omega_1,\ldots,\omega_M)$, $\omega_j\in\{0,1\}$, $j = 1,\ldots,M$, $0 < \bar{c} < 1$, and r is an integer such that $0 \le r \le m^\gamma-1$. The corresponding domains are of the form

$$G^r(\bar{\omega}) = \left\{ x \in K:\ 0 \le x_N \le g^r(x_1,\dots,x_{N-1},\bar{\omega}) \right\}.$$

Denote

$$\mathcal{T}_{jr} = \left\{ x \in K:\ g^r(x_1,\dots,x_{N-1},\bar{\omega}_{j0}) < x_N \le g^r(x_1,\dots,x_{N-1},\bar{\omega}_{j1}) \right\} =$$

$$= \left\{ x \in K:\ |t-b_j| < 1/(2m),\ r/m^{\gamma} < x_N \le r/m^{\gamma} + (\bar{c}/m^{\gamma})\phi(m(t-b_j)) \right\}$$

where, as in Theorem 2.6.4,

$$\bar{\omega}_{j0} = (\omega_1,\dots,\omega_{j-1},0,\omega_{j+1},\dots,\omega_M),$$

$$\bar{\omega}_{j1} = (\omega_1,\dots,\omega_{j-1},1,\omega_{j+1},\dots,\omega_M),$$

and $|t-b_j|$ is the maximum-component norm. Let D_m be the set of integers r such that $hm^{\gamma} < r < (1-h)m^{\gamma} - 1$. If $r \in D_m$ and $\bar{c}$ is small enough then the functions $g^r(t,\bar{\omega})$ for each $\bar{\omega}$ belong to the class $\Sigma_{N-1}(\gamma,L,h)$ (cf. Proposition 2.4.3).

For $\mathcal{X} = (X_1,\dots,X_n)$ fixed and for m large enough there exist a subset $D_{1m}=D_{1m}(\mathcal{X})$ of D_m and subsets $\Psi_r=\Psi_r(\mathcal{X})$ of $\{1,\dots,M\}$ such that

$$\text{card } D_{1m} \ge (9/16)(1/2-h)m^{\gamma}$$

$$\text{card } \Psi_r \ge M/2,\ r \in D_{1m}, \tag{5.2}$$

$$\text{card } U_{jr} \le 8/(1-2h),\ j \in \Psi_r,\ r \in D_{1m},$$

where $U_{jr}=\{i:\ X_i \in \mathcal{T}_{jr}\}$. Let us prove (5.2). Denote

$$\mathcal{S}_r = \left\{ x \in K:\ x_N \in [r/m^{\gamma},(r+1)/m^{\gamma}) \right\},\quad r = 0,\dots,m^{\gamma}-1.$$

There are at least $(3/4+h/2)m^{\gamma}$ integers r such that

$$\text{card } \{i:\ X_i \in \mathcal{S}_r\} \le 4n/((1-2h)m^{\gamma}). \tag{5.3}$$

Denote by D_{1m} the set of integers $r \in D_m$ that satisfy (5.3). Let $m^{\gamma} > 16/[3(1-2h)]$. Then

$$\text{card } D_{1m} \ge \left((3/4+h/2)m^{\gamma} + \text{card } D_m - m^{\gamma}\right)/2 > (9/16)(1/2-h)m^{\gamma}.$$

This is the first inequality in (5.2). Now, let

$$\Psi_r = \left\{ j: \text{ card } U_{jr} \le 8/(1-2h) \right\}.$$

Then the second and the third lines of (5.2) follow. In fact, assume that card $\Psi_r < M/2$. Then

$$\text{card } \{i: X_i \in \mathcal{S}_r\} \ge \text{card} \bigcup_{j \notin \Psi_r} U_{jr} \ge 8(M - \text{card } \Psi_r)/(1-2h) > 4M/(1-2h)$$

which contradicts (5.3).

Denote $E_{\mathcal{X}}$ the expectation with respect to $\mathcal{X}=(X_1,\dots,X_n)$. Similarly to (2.45)-(2.47) one gets

$$(5.4) \qquad \sup_{G \in \mathcal{G}_{0,N}} E_G(d_1(G,\hat{G}_n)) \ge \max_{r \in D_m} \max_{\bar{\omega} \in \Omega} E_{G^r(\bar{\omega})}(d_1(G^r(\bar{\omega}),\hat{G}_n)) \ge$$

$$\ge E_{\mathcal{X}}\left[(\text{card } D_m)^{-1} \sum_{r \in D_m} 2^{-M} \sum_{\bar{\omega} \in \Omega} \sum_{j=1}^{M} E_{G^r(\bar{\omega})}\left(d_1(G^r(\bar{\omega}) \cap \mathcal{T}_{jr}, \hat{G}_n \cap \mathcal{T}_{jr}) | \mathcal{X}\right)\right]$$

$$\ge (\text{card } D_m)^{-1} 2^{-M} E_{\mathcal{X}}\left[\sum_{r \in D_{1m}} \sum_{j \in \Psi_r} \left\{ \sum_{\bar{\omega} \in \Omega} E_{G^r(\bar{\omega})}\left(d_1(G^r(\bar{\omega}) \cap \mathcal{T}_{jr}, \hat{G}_n \cap \mathcal{T}_{jr}) | \mathcal{X}\right)\right\}\right]$$

where the difference from (2.45)-(2.47) is in the fact that we take also another maximum, over $r \in D_m$. As in (2.48), the expression in curly brackets in (5.4) can be represented as

$$\sum{}' \left[E_{G^r(\bar{\omega}_{j0})}\left(d_1(G^r(\bar{\omega}_{j0}) \cap \mathcal{T}_{jr}, \hat{G}_n \cap \mathcal{T}_{jr}) | \mathcal{X}\right) + \right.$$

$$+ E_{G^r(\bar{\omega}_{j1})}\left(d_1(G^r(\bar{\omega}_{j1})\cap\mathcal{I}_{jr},\hat{G}_n\cap\mathcal{I}_{jr})|\mathcal{X}\right)\Big]$$

where

$$\sum{}' = \sum_{\omega_1=0,1}\cdots\sum_{\omega_{j-1}=0,1}\ \sum_{\omega_{j+1}=0,1}\cdots\sum_{\omega_M=0,1} .$$

By the triangle inequality

$$d_1(G^r(\bar{\omega}_{j0})\cap\mathcal{I}_{jr},\hat{G}_n\cap\mathcal{I}_{jr}) + d_1(G^r(\bar{\omega}_{j1})\cap\mathcal{I}_{jr},\hat{G}_n\cap\mathcal{I}_{jr}) \geq \tag{5.5}$$

$$\geq d_1(G^r(\bar{\omega}_{j1})\cap\mathcal{I}_{jr},G^r(\bar{\omega}_{j0})\cap\mathcal{I}_{jr}) = \text{mes}\ \mathcal{I}_{jr} \geq c_1 n^{-1}$$

where $c_1>0$ is a constant. Let $P_{i,G}$ be the distribution of Y_i satisfying (3.1) or (5.1) in case when the image domain is G and X_i is fixed. It follows from (5.2), (5.4), (5.5) that for $j\in\Psi_r$, $r\in D_{1m}$

$$\Big[E_{G^r(\bar{\omega}_{j0})}\left(d_1(G^r(\bar{\omega}_{j0})\cap\mathcal{I}_{jr},\hat{G}_n\cap\mathcal{I}_{jr})|\mathcal{X}\right) + \tag{5.6}$$

$$+ E_{G^r(\bar{\omega}_{j1})}\left(d_1(G^r(\bar{\omega}_{j1})\cap\mathcal{I}_{jr},\hat{G}_n\cap\mathcal{I}_{jr})|\mathcal{X}\right)\Big] \geq$$

$$\geq (c_1 n^{-1})\int\min\left(\prod_{i\in U_{jr}} dP_{i,G^r(\bar{\omega}_{j0})},\ \prod_{i\in U_{jr}} dP_{i,G^r(\bar{\omega}_{j1})}\right) \geq$$

$$\geq (c_1 n^{-1}/2)\prod_{i\in U_{jr}}\left(1 - H^2\left(P_{i,G^r(\bar{\omega}_{j0})},P_{i,G^r(\bar{\omega}_{j1})}\right)/2\right)^2$$

where we used Le Cam's inequality (Corollary 2.3.6 of Chapter 2). Here $H(P,Q)$ is the Hellinger distance between probability measures P and Q. Now, for the model (3.1) with $f(x)=I\{x\in G\}$ we have

$$H^2\left(P_{i,G^r(\bar{\omega}_{j0})}, P_{i,G^r(\bar{\omega}_{j1})}\right) = H^2(1).$$

It follows from Assumption 3.3.5 and the last inequality in (5.2) that the last expression in (5.6) is greater than $c_2 n^{-1}$, $c_2>0$, if $j\in \Psi_r$, $r\in D_{1m}$. This together with (5.4) implies

$$\sup_{G\in\mathcal{G}_{0,N}} E_G(d_1(G,\hat{G}_n)) \geq$$

$$\geq (\text{card } D_m)^{-1}2^{-M}E_{\mathcal{X}}\left[\sum_{r\in D_{1m}} \sum_{j\in\Psi_r}\left\{\sum{}' c_2 n^{-1}\right\}\right].$$

This inequality and the first two inequalities in (5.2) entail the theorem. For the model (5.1) the argument is the same since it can be easily seen that for this model the squared Hellinger distance in (5.6) is also less than 2. □

5.1.2. THEOREM. Let (3.15) and Assumption 3.3.5 be satisfied and let the design $(X_1,\dots X_n)$, $X_i\in[0,1]^N$, $i=1,\dots,n$, be arbitrary but independent of $(\xi_1,\dots,\xi_n)$. Then for the model (3.1) with $f\in\Phi^0_{\infty,\gamma,N}$ we have

$$\liminf_{n\to\infty}\ \inf_{T_n}\ \sup_{f\in\Phi^0_{\beta,\gamma,N}} \psi_n^{-2}\text{MISE}(T_n,f) \geq C_0 > 0$$

where

$$\psi_n = \max\left\{(1/n)^{\gamma/(2(\gamma+N-1))},\ (1/n)^{\beta/(2\beta+N)}\right\},\quad \gamma\geq 1,\ 0<\beta<\infty,$$

(5.7)

$$\psi_n = (1/n)^{\gamma/(2(\gamma+N-1))},\quad \gamma\geq 1,\ \beta=\infty.$$

5.1.3. REMARK. The constants C_0 in Theorems 5.1.1 and 5.1.2 do not depend on the design.

Theorem 5.1.2 is reduced to Theorem 5.1.1 in the same way as Theorem 3.3.7 was reduced to Theorem 3.3.6. So, we do not give the proof here.

Let us describe the optimal edge estimator for N-dimensional binary boundary fragments. It is a generalization of the estimator $\hat{g}$ introduced in Section 4.1 for the two-dimensional case. We need some definitions of general character.

5.1.4. DEFINITION. Let Θ be a compact metric space with metric $\tilde{d}$, and let $\varepsilon>0$ be fixed. The set $\Theta' \subseteq \Theta$ is called ε-net on Θ in $\tilde{d}$-metric if for any $\vartheta \in \Theta$ there exists $\vartheta' \in \Theta'$ such that $\tilde{d}(\vartheta, \vartheta') < \varepsilon$.

Since Θ is compact there exists a finite ε-net on Θ for any $\varepsilon>0$. Denote $\mathcal{N}_\varepsilon$ a minimal ε-net, i.e. an ε-net with minimal number of elements. Note that $\mathcal{N}_\varepsilon$ is not necessarily uniquely defined.

Now we proceed to the definition of edge estimator. Consider the cubes Q_ℓ defined in the proof of Theorem 5.1.1, with edges of length 1/m. Set

$$m = m_n = \begin{cases} [n^{1/(\gamma+N-1)}] & \text{if } d=d_1, \\ [(n/\log n)^{1/(\gamma+N-1)}] & \text{if } d=d_\infty. \end{cases}$$

Define the corresponding slices on $K=[0,1]^N$:

$$A_\ell = \left\{ (x_1, \ldots, x_N) \in K:\ (x_1, \ldots, x_{N-1}) \in Q_\ell \right\}$$

In each cube Q_ℓ define the estimator $\hat{g}(x_1, \ldots, x_{N-1})$ as a polynomial $p(x_1, \ldots, x_{N-1})$ of degree $k=\lfloor\gamma\rfloor$ (k is the maximal integer such that $k<\gamma$) solving the following minimization problem

$$\min_{p \in \mathcal{P}_{N,k}} \sum_{X_i \in Q_\ell} \left(Y_i - I\{X_{i(N)} \le p(X_{i(1)}, \ldots, X_{i(N-1)})\} \right)^2$$

where $X_{i(j)}$ is the jth coordinate of X_i, and $\mathcal{P}_{N,k}$ is the finite set of polynomials defined as follows:

$$\mathcal{P}_{N,k} = \{p(x_1,\ldots,x_{N-1})=p(x_1,\ldots,x_{N-1},\theta),\ \theta\in\mathcal{N}_\varepsilon\}.$$

Here θ is the vector of coefficients of a polynomial $p(x_1,\ldots,x_{N-1})=p(x_1,\ldots,x_{N-1},\theta)$ of degree k and $\mathcal{N}_\varepsilon$ is a minimal ε-net (with $\varepsilon \sim m_n^{-\gamma}$) on the set

$$\Theta = \{\theta:\ h \le p(x_1,\ldots,x_{N-1},\theta) \le 1-h,\ \forall\ (x_1,\ldots,x_{N-1})\in Q_\ell\}$$

in the Euclidean metric.

Denote G_n^* the closure of the set

$$\Big\{(x_1,\ldots,x_N):\ 0 \le x_N \le \hat{g}(x_1,\ldots,x_{N-1})\Big\}. \tag{5.8}$$

Convergence rate of $\hat{g}$ is studied similarly to the two-dimensional case (see Sections 4.2 and 4.3). Here we state the corresponding result in most general form allowing the noise to be non-Gaussian and nonidentically distributed. We use the following assumption (it is identical to 1.9.7; we reproduce it here for convenience only).

5.1.5. ASSUMPTION. Random variables ξ_i in (3.1) are conditionally independent for fixed $\mathcal{X}=(X_1,\ldots,X_n)$, and

$$E(\xi_i|\mathcal{X}) = 0,$$

$$E(\exp(z\xi_i)|\mathcal{X}) \le \exp(Hz^2),\ i=1,\ldots,n,$$

for any real z, where $H>0$ does not depend on $i,\mathcal{X},f$.

For example, Assumption 5.1.5 is satisfied under the multiplicative model (5.1) (see (1.48)).

5.1.6. THEOREM. Let Assumption 5.1.5 hold and let the design satisfy (B1). Then the estimator G_n^* which is the closure of the set (5.8) satisfies

$$\sup_{G\in\mathcal{G}_{0,N}} E_G(\psi_n^{-q} d^q(G,G_n^*)) \le C_1,\ q>0,$$

where C_1 is a constant depending on q, and

(i) $\psi_n=(1/n)^{\gamma/(\gamma+N-1)}$ if d is the measure of symmetric difference d_1;

(ii) $\psi_n=(\log n/n)^{\gamma/(\gamma+N-1)}$ if d is the Hausdorff distance d_∞.

Proof of the Theorem is essentially the same as the proof of Theorem 4.1.1, with the change of notation corresponding to N-dimensional case. Convergence for arbitrary q>0 follows from exponential decrease of the tails of noise distribution as guaranteed by Assumption 5.1.5 (see Remark 4.3.2). Note that under Assumption 5.1.5 $\hat{g}$ is not the MLE in a slice but rather the LSE. This does not essentially change the proof (cf. Lemma 4.4.3).

5.1.7. REMARK. Theorem 5.1.6 remains valid if the variables ξ_i depend on n provided the constant H in Assumption 5.1.5 is independent of n.

Finally, let us proceed to image estimation for N-dimensional grey-scale boundary fragments. Consider only the random uniform design (B1). Construction of optimal image and edge estimators follows the two-step procedure of Section 4.4. Let $p_{n,\Delta,N}$ be the N-dimensional analogue to $p_{n,\Delta}$ of Section 4.4. Set

$$\hat{f}_n(x) = p_{n,\Delta,N}(x) I\{0 \le p_{n,\Delta,N}(x) \le 1\}. \tag{5.9}$$

5.1.8. THEOREM. Let Assumption 5.1.5 hold and let the design satisfy (B1). Assume that for the edge estimation step the bin width is $\delta_n=n^{-1/(\gamma+N-1)}$, and for the smoothing step the window width is $h_n=n^{-1/(2\beta+N)}$. Then for the image estimator (5.9) the upper bound (3.8) holds with $\Phi=\Phi^0_{\beta,\gamma,N}$ and ψ_n as in (5.7).

As a conclusion, indicate the main points of this section:

(i) It is possible to find optimal image and edge estimators in dimension N using the same ideas as in two-dimensional case.

(ii) The A-minimax rate of convergence in N-dimensional edge estimation for boundary fragments in distance d_1 is $(1/n)^{\gamma/(\gamma+N-1)}$, and the minimax rate in N-dimensional image estimation for grey-scale boundary fragments is given by (5.7).

For the Hausdorff distance d_∞ we have given only the upper bound, with the rate $(\log n/n)^{\gamma/(\gamma+N-1)}$. Surely, it is not difficult to show that the I-lower rate of convergence is the same, and thus our estimator is I-optimal for the classes of N-dimensional boundary fragments in Hausdorff metric. This can be done by a straightforward extension of Theorem 3.3.2 to N-dimensional case.

5.2. GENERAL DOMAINS IN HIGH DIMENSIONS: A SIMPLE AND ROUGH ESTIMATOR

The results of Chapters 3, 4 and Section 5.1 cover only the case of boundary fragments. Here we consider an approach to estimation of N-dimensional images with general and, perhaps, bad domains. This approach is based on rough estimators with the rates of convergence that do not get faster as the edge smoothness increases. To define the general class of domains we need some notation.

5.2.1. DEFINITION. Let $0<a_1<1$, $a_2>0$. A set G satisfies the (a_1,a_2)-cone condition iff for any point $x\in G$ there exists a direction $u\in R^N$, $|u|=1$, such that the cone

$$\{z\in R^N: |z-x|\le a_2, (z-x)^T u\ge a_1|z-x|\}$$

is contained in G.

Let $\mathcal{G}_N$ be the set of domains G satisfying

5.2.2. ASSUMPTION. For $G\in\mathcal{G}_N$:

(i) The edge $\Gamma=\partial G$ is piecewise Lipschitz with the number of pieces and the Lipschitz constant bounded uniformly on $\mathcal{G}_N$.

(ii) The domain G and the background $\bar{G}$ satisfy the (a_1,a_2)-cone condition where a_1,a_2 are the same for all $G\in\mathcal{G}_N$.

The class $\mathcal{G}_N$ is rather general. For example, the sets $G\in\mathcal{G}_N$ are not necessarily connected.

Let $0<\beta<\infty$. Define the class of images Φ_β as the set of all functions f on $K=[0,1]^N$ such that

$$f(x)=f_0(x)I\{x\in\bar{G}\}+f_1(x)I\{x\in G\}$$

where $G\in\mathcal{G}_N$, f_0,f_1 satisfy (3.2), and $f_0,f_1\in\Sigma_N(\beta,L_1)$, $0<L_1<\infty$.

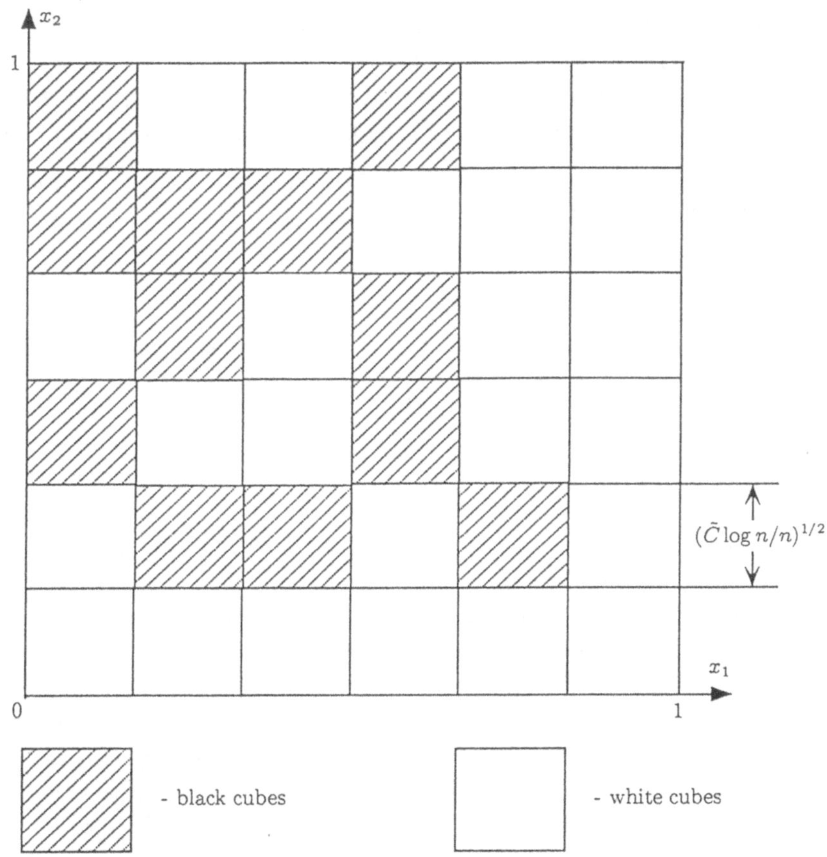

Fig. 5.1. To the definition of G_n^{**} (two-dimensional case).

Consider a two-step image estimator. The difference from estimators of Section 4.4 is essentially in the first step, i.e. in the way of classifying into white and black parts. Now it is a kind of domain estimation (not edge estimation since the edges can be of a rather complex structure for $G\in\mathcal{G}_N$). The second step is smoothing based on the direct N-dimensional extension of the method described in Section 4.4.4.

5.2.3. CLASSIFICATION STEP. Let $\tilde{Q}_\ell$, $\ell=1,\dots,m^N$, be disjoint cubes with edges of length $1/m$ that form a partition of K. Let $m=m_n$ be an integer such that $1/m_n \sim (\tilde{C}(\log n)/n)^{1/N}$, $n\to\infty$, $\tilde{C}>0$. For all cubes $\tilde{Q}_\ell$ calculate the medians

$$\text{med}_\ell = \text{med}\{Y_i : X_i \in \tilde{Q}_\ell\}.$$

Classify the cubes into black and white ones according to the following rule. Decide that $\tilde{Q}_\ell$ is black if $\text{med}_\ell < (a+b)/2$ and that $\tilde{Q}_\ell$ is white if $\text{med}_\ell \geq (a+b)/2$. Denote

$$G_n^{**} = \bigcup_\ell \{ \tilde{Q}_\ell : \tilde{Q}_\ell \text{ is white } \}.$$

The estimator G_n^{**} of the domain G has a structure of "irregular chess-board" shown in Fig. 5.1.

Note that the number of design points in each cube is roughly $\tilde{C}\log n$ under either of the designs (A1), (A2), (B1), (B2) (here and later in this section (A1), (A2), (B2) are direct N-dimensional generalizations of the corresponding two-dimensional designs). Thus, in view of the exponential error bound for deviations of the median the classification error is decreasing faster than any prescribed power of n^{-1} for $\tilde{C}$ large enough.

5.2.4. SMOOTHING STEP. Use the method of Section 4.4.4 with N-dimensional data and set there $G_n^* = G_n^{**}$, $h_n = n^{-1/(2\beta^*+N)}$,

$$\beta^* = \min\{\beta, N/(2(N-1))\}.$$

5.2.5. THEOREM. Assume that the design is of type (A1), (A2), (B1), or (B2). Let ξ_i be i.i.d. random variables with a symmetric distribution function F such that $F(t)>1/2$, $t>0$. Then for $\tilde{C}>0$ large enough

$$\sup_{f\in\Phi_\beta} \mathrm{MISE}(\hat{f}_n, f) = O(\max\{n^{-2\beta/(2\beta+N)}, (\log n/n)^{1/N}\}), \beta>0,$$

where $\hat{f}_n$ is the two-step estimator defined by 5.2.3 and 5.2.4.

5.2.6. REMARK. The choice of $\tilde{C}$ depends only on the noise distribution F.

Proof of this theorem in a slightly different case is given in Tsybakov (1989). Here we prove a simpler version of Theorem 5.2.5 which corresponds to edge estimation problem, i.e. to the problem of estimation of domains G belonging to the class $\mathcal{G}_N$. This problem is a special case of the previous one, with $f_0(x)\equiv 0$, $f_1(x)\equiv 1$, a=0, b=1. The estimator G_n^{**} of G is defined as a union of all white cubes $\tilde{Q}_\ell$ where $\tilde{Q}_\ell$ is white if $\mathrm{med}_\ell>1/2$. For this estimator we have the following.

5.2.7. THEOREM. Assume that the design is of type (A1), (A2) or (B2). Let ξ_i be i.i.d. random variables with symmetric distribution function F such that $F(1/2) > 1/2$. Then for $\tilde{C} > 0$ large enough

$$\sup_{G\in\mathcal{G}_N} E_G(d(G,G_n^{**})) = O((\log n/n)^{1/N}), \ n \to \infty,$$

with $d=d_1$ and $d=d_\infty$.

5.2.8. REMARK. As follows from the proof, condition 5.2.2(ii) of the definition of $\mathcal{G}_N$ is not necessary in this theorem.

To prove the theorem we need the next lemma.

5.2.9. LEMMA. If $G\in\mathcal{G}_N$ then under the conditions of Theorem 5.2.7 there exists a constant $\varepsilon > 0$ such that

$$P_G\left\{\{\tilde{Q}_\ell\cap G=\emptyset, \ \tilde{Q}_\ell \text{ is white}\} \cap \{\tilde{Q}_\ell\subset G, \ \tilde{Q}_\ell \text{ is black}\}\right\} \le 2\exp(-\tilde{C}\varepsilon^2\log n),$$

$$\ell=1,\dots,m_n^N.$$

<u>Proof</u>. Let $n_\ell=\mathrm{card}\{i:X_i\in\tilde{Q}_\ell\}$. For the designs (A1),(A2),(B2) we have if n is large enough:

$$n_\ell \geq nm_n^{-N} \geq \frac{\tilde{C}}{2} \log n$$

with probability 1.

Let $F_{n\ell}(t)$, $t\in R^1$, be the empirical distribution function of Y_i's such that $X_i \in \tilde{Q}_\ell$. Denote

$$F_\ell(t) = \begin{cases} F(t) \text{ if } \tilde{Q}_\ell \cap G = \emptyset, \\ F(t-1) \text{ if } \tilde{Q}_\ell \subseteq G. \end{cases}$$

Now we use Hoeffding's(1963) inequality: let $\tilde{\eta}_1, \ldots \tilde{\eta}_n$ be i.i.d. random variables such that $c_3 \leq \tilde{\eta}_i \leq c_4$, $i=1,\ldots,n$, then

$$P\left(\frac{1}{n}\sum_{i=1}^{n} (\tilde{\eta}_i - E(\tilde{\eta}_i)) \geq \varepsilon\right) \leq \exp\left(-2n\varepsilon^2/(c_4-c_3)^2\right)$$

for any $\varepsilon>0$.

If $\tilde{Q}_\ell \cap G=\emptyset$ then $F(1/2) = 1/2+\varepsilon$ for some $\varepsilon>0$ by assumption of the theorem. Thus,

$$P_G\{\tilde{Q}_\ell \cap G=\emptyset,\ \tilde{Q}_\ell \text{ is white}\} \leq P_G\{F(1/2) = 1/2+\varepsilon,\ \mathrm{med}_\ell > 1/2\} \leq$$

$$\leq P_G\{F(1/2) = 1/2+\varepsilon,\ F_{n\ell}(1/2) \leq 1/2\} \leq P_G\{F_\ell(1/2)-F_{n\ell}(1/2) \geq \varepsilon\}=$$

$$= E_{n_\ell}(P_G\{F_\ell(1/2)-F_{n\ell}(1/2) \geq \varepsilon | n_\ell\}) \leq E_G(\exp(-2n_\ell\varepsilon^2) | n_\ell) \leq$$

$$\leq \exp(-\tilde{C}\varepsilon^2 \log n)$$

by Hoeffding's inequality. The probability

$$P_G\{\tilde{Q}_\ell \subset G,\ \tilde{Q}_\ell \text{ is black}\}$$

is estimated similarly. □

Proof of Theorem 5.2.7. Denote $Q = \bigcup_\ell \{\tilde{Q}_\ell:\ \partial G \cap \tilde{Q}_\ell \neq \emptyset\}$. It follows from Assumption 5.2.2 (i) that the δ-neighborhood of the edge ∂G has Lebesgue measure $O(\delta)$ for δ small enough. Since Q is embedded into $N^{1/2}/m_n$-neighborhood of ∂G we have

$$d(G, G \cap Q) = O(m_n^{-1}),\ n \to \infty, \tag{5.10}$$

for $d=d_1,d_\infty$.

Introduce the random event

$$\mathcal{A} = \left\{ \text{all } \tilde{Q}_\ell \subset G \text{ are white and all } \tilde{Q}_\ell \text{ such that } \tilde{Q}_\ell \cap G=\emptyset \text{ are black} \right\}.$$

Using (5.10) we get for $d=d_1,d_\infty$

$$P_G\{m_n d(G,G_n^{**}) \geq t\} \leq$$

$$\leq P_G\left\{ \{m_n d(G,\bigcup_\ell\{\tilde{Q}_\ell: \tilde{Q}_\ell \text{ is white}\}) \geq t\} \cap \mathcal{A} \right\} + P_G\{\bar{\mathcal{A}}\} \leq$$

$$\leq P_G\{m_n d(G,G\cup Q) \geq t\} + P_G\{\bar{\mathcal{A}}\} \leq P_G\{\bar{\mathcal{A}}\}$$

for $t \geq t_0 > 0$.

Lemma 5.2.9 entails

$$P_G\{\bar{\mathcal{A}}\} \leq 2m_n^N \exp(-\tilde{C}\varepsilon^2 \log n). \tag{5.11}$$

Now, since $d(G,G_n^{**}) \leq N^{1/2}$,

$$E_G(m_n d(G,G_n^{**})) \leq \int_0^{m_n N^{1/2}} P_G\{m_n d(G,G_n^{**}) \geq t\}dt \leq$$

$$\leq t_0 + \int_{t_0}^{m_n N^{1/2}} P_G\{m_n d(G,G_n^{**}) \geq t\}dt \leq t_0 + m_n N^{1/2} P_G\{\bar{\mathcal{A}}\}.$$

Choosing $\tilde{C} > (1+1/N)/\varepsilon^2$ and applying (5.11) we get the theorem. □

5.3. OPTIMAL ESTIMATORS FOR GENERAL DOMAINS IN TWO DIMENSIONS

The rough estimator introduced in Section 5.2 has poor rate of convergence. Comparison with Theorems 5.1.6 and 5.1.8 shows that this rate is worse than the minimax rate for boundary fragments with the smoothness parameter $\gamma>1$. Thus, one may suppose that for general domains the rate of convergence becomes worse than for boundary fragments. In fact this is not true. The rate depends only on γ, the smoothness parameter of the boundary, and does not depend on the topological properties of the boundary. In this section we show that for one class of general domains in two dimensions the minimax rates are the same as for boundary fragments. We give an explicit construction of optimal estimators for general domains with rather smooth boundaries using the estimator of Section 5.2 as a pilot one. The optimality of convergence rates is proved in the Hausdorff metric. Although the estimator is described explicitly, it is mainly of theoretical value.

Define the class $\mathcal{G}$ of domains G by the following conditions.

(i) All the domains from $\mathcal{G}$ are closed connected compacts in $K=[0,1]^2$. Distances between the edges $\Gamma=\partial G$ and the boundary of K are bounded away from zero uniformly on $\mathcal{G}$.

(ii) The length $|\Gamma|$ is bounded uniformly on $\mathcal{G}$.

(iii) Under the natural parameterization the curve Γ is represented as $(x_1(s),x_2(s))$, $0\le s\le|\Gamma|$, and the functions $x_1(s/|\Gamma|)$, $x_2(s/|\Gamma|)$ belong to the class $\Sigma(\gamma,L,h)$ with some fixed γ,L,h.

(iv) The curve Γ has no singular points, i.e. the function

$$\dot{x}_1^2(s)+\dot{x}_2^2(s)$$

is bounded away from zero for all $0\le s\le|\Gamma|$.

(v) $\gamma\ge2$, so that the minimal radius of curvature r_0 is bounded away from 0 uniformly over the domains $G\in\mathcal{G}$; intersection of the edge $\Gamma=\partial G$ with any ball of radius r_0 is a connected or empty set.

Some comments are in order here. Condition (v) rules out the edges having self-intersections or "almost" self-intersections appearing when the distances between some disjoint parts of Γ are small (less than r_0, see Fig.5.2(b)). The situation shown in Fig. 5.2(a) is excluded by condition (v). This means that the r_0-neighborhood of Γ is topologically equivalent to a ring.

By conditions (i)-(v) the edge Γ is strictly separated from the boundary of K and the curvature of Γ is bounded. Hence it is

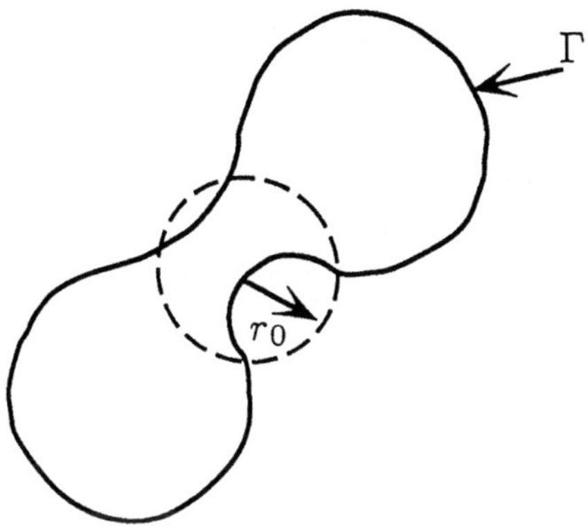

(a)

intersection of Γ with the circle of radius r_0
is not connected

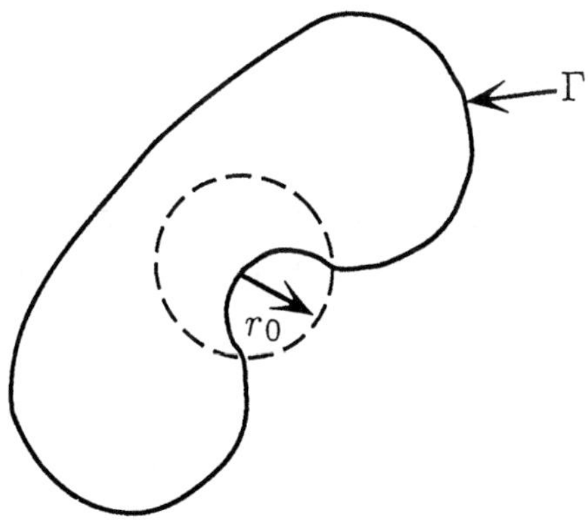

(b)

intersection of Γ with the circle of radius r_0
is connected

Fig. 5.2.

possible to cover a small neighborhood of Γ by finitely many overlapping rectangles. In each of these rectangles the problem can be reduced to the special case of boundary fragment by an appropriate rescaling.

Optimal estimator is defined in two steps. For this purpose we split the whole sample $((X_1,Y_1),\ldots,(X_n,Y_n))$ into two independent subsamples containing observations with odd and even numbers respectively. For the first step define a rough pilot estimator $G_n^{(0)}$ having a finite error and a smooth edge. What we need is the following:

$$\lim_{n\to\infty} \sup_{G\in\mathcal{G}} P_G(d_\infty(G,G_n^{(0)})>r_1) = 0 \tag{5.12}$$

where r_1 is several times smaller than r_0, say $r_1=r_0/100$.

The following gives an example of $G_n^{(0)}$ satisfying (5.12). Let G_n^{**} be the estimator of Section 5.2 (with N=2) for the first subsample. Denote by $\tilde{G}_n$ the largest connected component of G_n^{**}, i.e. the component having the largest Lebesgue measure. Let $G_n^{(0)}$ be the element of $\mathcal{G}$ closest to $\tilde{G}_n$, i.e. $G_n^{(0)} = \underset{G\in\mathcal{G}}{\operatorname{argmin}}\, d_\infty(G,\tilde{G}_n)$. Clearly, $G_n^{(0)}$ exists since $\mathcal{G}$ is compact.

5.3.1. PROPOSITION. Let $\mathcal{G}$ be the class of domains satisfying (i)-(v) of this section and let the assumptions of Theorem 5.2.7 hold. Then (5.12) holds for an arbitrary constant $r_1>0$.

<u>Proof</u>. Since $\mathcal{G}\subset\mathcal{G}_2$ we have by Theorem 5.2.7 that for any $\varepsilon>0$

$$\lim_{n\to\infty} \sup_{G\in\mathcal{G}} P_G(d_\infty(G,G_n^{**})>(\varepsilon/2)(\log n/n)^{1/4}) = 0.$$

If

$$d_\infty(G,G_n^{**})\le(\varepsilon/2)(\log n/n)^{1/4} \tag{5.13}$$

then also

$$d_\infty(G,\tilde{G}_n)\le(\varepsilon/2)(\log n/n)^{1/4} \tag{5.14}$$

for n large enough. In fact, (5.13) entails that G differs from G_n^{**} only in some part of the stripe of width $\varepsilon(\log n/n)^{1/4}$ around $\Gamma=\partial G$. If n is large enough then the Lebesgue measure of the stripe is smaller than that of the inner part of G (the common part of G and G_n^{**} that does not intersect the stripe). Hence the largest connected component of G_n^{**} contains the inner part of G, and thus (5.14) holds.

Now, (5.14) and the definition of $G_n^{(0)}$ imply

$$d_\infty(G,G_n^{(0)}) \le d_\infty(G,\tilde{G}_n) + d_\infty(\tilde{G}_n,G_n^{(0)}) \le 2d_\infty(G,\tilde{G}_n) \le \varepsilon(\log n/n)^{1/4} \le r_1$$

for n large enough. □

Describe now an auxiliary geometric construction that will be used for the second step of our procedure.

5.3.2. EXTRACTION OF BOUNDARY FRAGMENTS. Let $\Gamma_n^{(0)}=\partial G_n^{(0)}$ be the edge of $G_n^{(0)}$. Choose some r_1-net $\{z_1,\dots,z_J\}$ on $\Gamma_n^{(0)}$ in the Euclidean metric. Points z_j as well as their number J are random but J has a nonrandom upper bound. Cover each point z_j by a rectangle R_j centered at z_j. Suppose that $d_\infty(G,G_n^{(0)}) \le r_1$ (this holds with probability close to 1 by (5.12)). Then the edge Γ is covered by the union of sets R_j if the size of R_j is larger than r_1 and each R_j is contained in the r_0-neighborhood of $\Gamma_n^{(0)}$. To satisfy these requirements assume that each rectangle has the sides of length $5r_1$ and $10r_1$ and that it is oriented according to the following rule. Let α_j be the angle between the x_2-axis and the normal direction to $\Gamma_n^{(0)}$ at point z_j. If $\alpha_j \le \pi/4$ then the longer side of the rectangle R_j is parallel to the x_2-axis, otherwise it is parallel to the x_1-axis.

For the second step, rescale the rectangles R_j to transform them into squares and in each of them apply that version of edge estimation procedure of Section 4.1 which is optimal with respect to the Hausdorff distance. Denote by $G_n^{(j)}$ the resulting estimators of $G \cap R_j$, $j=1,\dots,J$, based on the second subsample. With probability tending to 1 as $n \to \infty$ the set $W=\bigcup_{j=1}^{J} G_n^{(j)}$ covers a small neighborhood of Γ and its complement $\overline{W}=K\backslash W$ contains exactly

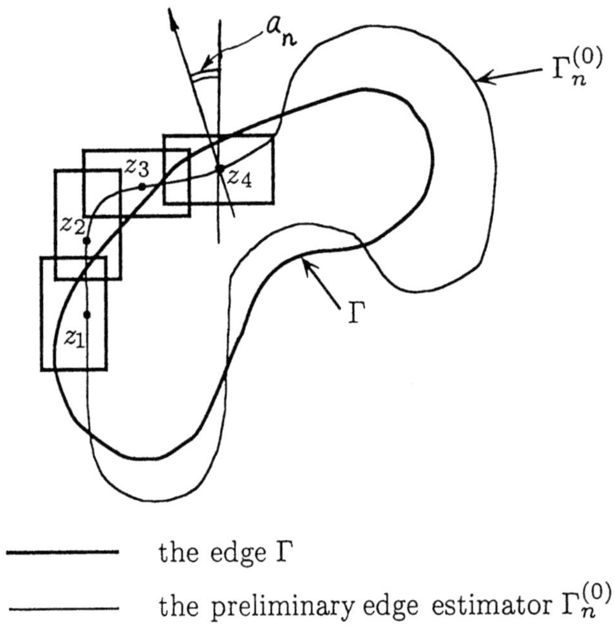

Fig. 5.3. Extraction of boundary fragments.

two disjoint connected components. Denote by Q the component which is separated from the boundary of K (Q is the inner polygon of W). Finally, define the edge estimator

$$G_n^* = \begin{cases} Q \cup W \text{ if } \overline{W} \text{ has exactly 2 disjoint connected components,} \\ W \text{ otherwise.} \end{cases}$$

It is clear from the above argument that G_n^* is optimal with respect to the Hausdorff distance. The result is stated as follows.

5.3.3. THEOREM. Let $\mathcal{G}$ be the class of domains satisfying the conditions (i)-(v) of this section. Then the normalized Hausdorff distance $\psi_n^{-1} d_\infty(G, G_n^*)$ is bounded in probability P_G uniformly over $G \in \mathcal{G}$, i.e.

$$\lim_{C\to\infty} \limsup_{n\to\infty} \sup_{G\in\mathcal{G}} P_G(\psi_n^{-1} d_\infty(G, G_n^*) > C) = 0$$

where $\psi_n = (\log n/n)^{\gamma/(\gamma+1)}$.

5.4. DUDLEY'S CLASSES OF DOMAINS

The rest of this chapter is devoted to the construction of optimal edge estimators for general domains in dimensions $N \ge 2$. We start with the definition of general smoothness classes of domains on $[0,1]^N$, $N \ge 2$, which include, in particular, the classes $\mathcal{G}$ considered in Section 5.3. The definition is a slight modification of that of Dudley (1974). It is rather sophisticated and requires some preliminary definitions. We give them later in this section. Now we would like to describe the idea of the construction qualitatively in two-dimensional case.

Assume that we have a closed curve g of smoothness γ on the plane R^2. We assign to this curve a domain G(g) following the next rule:

(a) Remove the curve from the plane. Then we are left with the union of disjoint open sets. Some of them which were "inside" the curve are bounded since the curve is closed. Also there is an outer unbounded set.

(b) Consider the union of all bounded sets that we obtained. Take the closure of this union. Denote it by $G^o(g)$.

(c) Define the domain G(g) as the intersection of $G^o(g)$ with $K=[0,1]^2$.

The class of domains G(g) defined this way for all closed curves g with given smoothness γ is what we call Dudley's class.

Note that this qualitative construction was given for two-dimensional case. For N>2 dimensions it should be suitably modified.

Now we proceed to the definitions.

First, consider the classes of smooth surfaces in R^N, i.e. the smooth maps of the unit ball $B^{N-1}=\{x\in R^{N-1}: |x|<1\}$ into R^N. For our purposes it would be more convenient to define the smooth maps on the unit sphere $S^{N-1}=\{s\in R^N: |s|=1\}$ rather than on the ball B^{N-1}. To make this change we introduce infinitely many times differentiable isomorphisms $\mathfrak{J}_j: B^{N-1}\to V_j$, $j=1,\dots,J$, of a unit ball onto the sets $V_j\subset S^{N-1}$ such that $\{V_j\}$ is a covering of the unit sphere: $\bigcup_j V_j = S^{N-1}$. The sets V_j, the isomorphisms $\mathfrak{J}_j$ and the number J are assumed to be fixed.

The class of surfaces having the "smoothness parameter" $\gamma>0$ and the "Lipschitz constant" L>0 is defined as

$$\Sigma S_N(\gamma,L) = \Big\{ g=(g_1,\dots,g_N) \mid g_i: S^{N-1}\to R^1,$$
$$g_i(\mathfrak{J}_j(\cdot))\in\tilde{\Sigma}_{N-1}(\gamma,L),\ j=1,\dots,J,\ i=1,\dots,N \Big\}.$$

Here $\tilde{\Sigma}_{N-1}(\gamma,L)$ is the smoothness class of functions defined in the same way as $\Sigma_{N-1}(\gamma,L)$ (see Section 2.7.6) but containing the functions on the ball B^{N-1}, not on the cube $[0,1]^{N-1}$.Note that the class ΣS_N is rather large (it contains, for example, surfaces with self-intersections).

For each surface $g\in\Sigma S_N(\gamma,L)$ introduce a domain $G^o(g)$ such that the set

$$\text{range}(g) = \{g(s):\ s\in S^{N-1}\}$$

or some parts of it form the boundary of $G^o(g)$. The definition of

$G^{o}(g)$ is based on the notion of homotopy. Let g and $\tilde{g}$ be two functions on S^{N-1} with values in $T\subseteq R^{N}$. Then g and $\tilde{g}$ are called <u>homotopic</u> iff there exists a continuous mapping F from $[0,1]\times S^{N-1}$ into T such that $F(0,\cdot)=g(\cdot)$ and $F(1,\cdot)=\tilde{g}(\cdot)$.

5.4.1. DEFINITION. The set $G^{o}(g)$ is defined as the closure of $G_{int}(g)$. The set $G_{int}(g)$ consists of all $x \notin \text{range}(g)$ such that in $T=R^{N}\setminus\{x\}$ the function g is not homotopic to any constant map of S^{N-1} into a point $x_0 \neq x$.

Roughly speaking, $g(s)$ is homotopic to a constant map $\tilde{g}(s)\equiv x_0 \neq x$ in $R^{N}\setminus\{x\}$ if it is possible to shrink the surface $g(s)$ into the point x_0 by a continuous transformation in such a way that the "intermediate" surfaces do not pass through x. All x that satisfy this property are not in $G_{int}(g)$. Let us give some examples.

5.4.2. EXAMPLES. (i) If range(g) is a sphere inside $[0,1]^{N}$ then $G_{int}(g)$ is the open ball bounded by this sphere and $G^{o}(g)$ is the closed ball.

(ii) See Fig. 5.4. Range(g) is the smooth curve with self-intersection shown in Fig. 5.4(a), $G_{int}(g) = A_1 \cup A_2$ where A_1 and A_2 are open sets in Fig. 5.4(b), and $G^{o}(g)$ is the set bounded by the solid line in Fig. 5.4(b). This example shows that although g is smooth the boundary of the set $G^{o}(g)$ is not necessarily smooth.

(iii) See Fig. 5.5. Range(g) is the smooth curve with many self-intersections shown in Fig. 5.5(a), $G_{int}(g) = \bigcup_{i=1}^{6} A_i$ where A_i are open sets in Fig. 5.5(b), and $G^{o}(g)$ is the union of two disjoint closed sets bounded by the solid line in Fig. 5.5(b). This example shows that $G^{o}(g)$ is not necessarily connected.

Now we are able to give the main definition of this section.

5.4.2. DEFINITION. The class of domains

$$\tilde{\mathcal{G}} = \tilde{\mathcal{G}}(N,\gamma,L) = \left\{G(g)=G^{o}(g)\cap[0,1]^{N}:\ g\in\Sigma S_N(\gamma,L)\right\}$$

is called <u>Dudley's class</u> of domains.

In the following we study the problem of estimating the set $G\in\tilde{\mathcal{G}}$ from observations satisfying the multiplicative model (5.1).

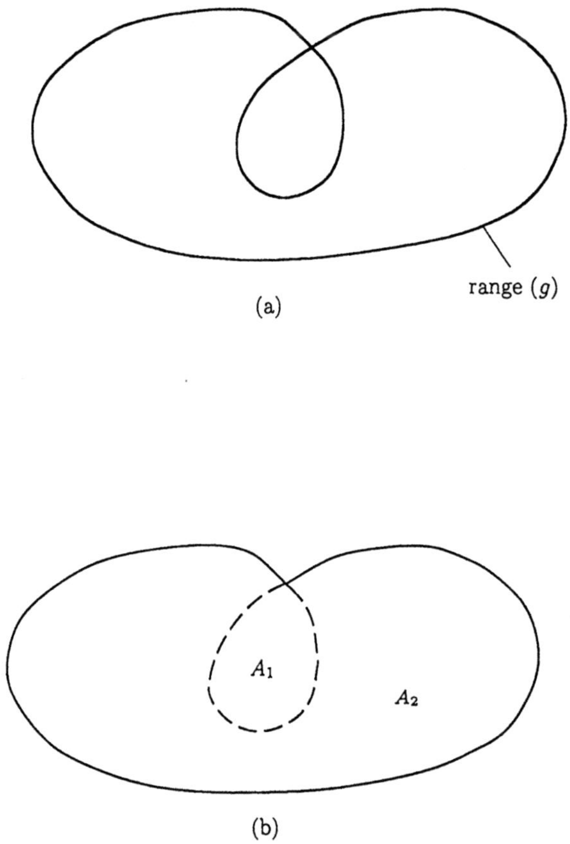

Fig. 5.4.

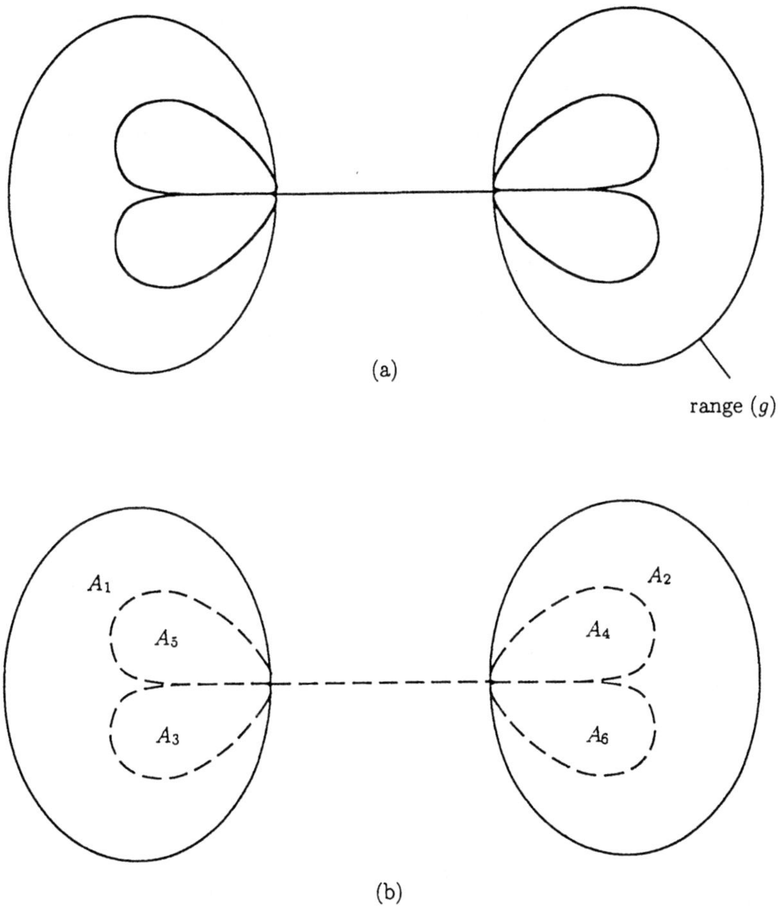

Fig. 5.5.

5.5. MAXIMUM LIKELIHOOD ESTIMATION ON ε-NET

The study of optimal edge estimators for the Dudley's classes of domains will require some general statistical tools, namely, the properties of the maximum likelihood estimator (MLE) on ε-net. Here we give the necessary definitions and a lemma.

Let Θ be a compact metric space with metric d, and let ε be a positive number.

5.5.1. DEFINITION. The ε-<u>entropy</u> $\mathcal{H}(\Theta,d,\varepsilon)$ of Θ in d-metric is the logarithm of minimal number of balls of d-radius less than ε necessary to cover Θ, i.e.

$$\mathcal{H}(\Theta,d,\varepsilon) = \log(\operatorname{card} \mathcal{N}_\varepsilon).$$

Here $\mathcal{N}_\varepsilon$ is a minimal ε-net on Θ in d-metric (see Definition 5.1.4).

Consider a general statistical model as defined in Section 2.1. Let $p(X^{(n)},\vartheta)$, $\vartheta\in\Theta$, be the distribution density for this model with respect to some measure ν.

5.5.2. DEFINITION. The estimator

$$\vartheta_n^* = \operatorname*{argmax}_{\theta\in\mathcal{N}_\varepsilon} p(X^{(n)},\vartheta)$$

of parameter ϑ is called <u>maximum likelihood estimator on ε-net</u>.

Assume that the density $p(X^{(n)},\vartheta)$ satisfies

$$(5.15) \qquad \sup_{\substack{\tilde\vartheta,\vartheta':d(\tilde\vartheta,\vartheta)\le\varepsilon,\\ d(\vartheta,\vartheta')\ge\varepsilon t}} P_\vartheta^{(n)}\Big(p(X^{(n)},\vartheta') > p(X^{(n)},\tilde\vartheta) \Big) \le$$

$$\le \exp(-c_d n(\varepsilon t)^\alpha), \quad t\ge t_0,$$

for every $\varepsilon>0$, where $c_d,\alpha > 0$, $t_0>1$ are constants that do not depend on ϑ.

If (5.15) holds then there exists a choice of ε that guarantees the best convergence rate in d-metric within the class of MLE's on ε-nets. This optimal choice of ε (which also equals to the best convergence rate) is roughly given as a solution of

$$\mathcal{H}(\Theta, d, \varepsilon) \sim n\varepsilon^{\alpha}.$$

The precise statement is presented in the following lemma which is a modification of a result in Birgé (1983).

5.5.3. LEMMA. Assume that (5.15) holds and $\varepsilon^*>0$ is such that

$$(5.16) \qquad \mathcal{H}(\Theta, d, \varepsilon^*) \le \tfrac{1}{2} c_d n (\varepsilon^*)^{\alpha} t^{\alpha}, \qquad t \ge t_1,$$

for some $t_1>0$. Then

$$(5.17) \qquad \sup_{\vartheta\in\Theta} E_{\vartheta}^{(n)}\Big[(\varepsilon^*)^{-q} d^q(\vartheta_n^*, \vartheta)\Big] \le \max(t_0, t_1) +$$

$$+ (D/\varepsilon^*)^q \exp\Big(-\tfrac{1}{2} c_d n (\varepsilon^*)^{\alpha}\Big), \quad \forall\, q>0,$$

where ϑ_n^* is the MLE on ε^*-net, and D is the diameter of Θ:

$$D = \sup_{\vartheta, \vartheta' \in \Theta} d(\vartheta, \vartheta').$$

5.5.4. REMARK. The values ε^* satisfying (5.16) always exist since the ε-entropy $\mathcal{H}(\Theta, d, \varepsilon)$ is a monotone non-increasing function of ε. Of course, the sharpest upper bound (5.17) is obtained if ε^* is a solution of the equation $\mathcal{H}(\Theta, d, \varepsilon) = \frac{1}{2} c_d n \varepsilon^{\alpha} t_1^{\alpha}$ (in cases where such a solution exists).

Proof. If ϑ_n^* is MLE on ε-net then for any $t>0$

$$P_{\vartheta}^{(n)}(d(\vartheta_n^*, \vartheta) \ge \varepsilon t) \le$$

$$\le P_{\vartheta}^{(n)}\Big(\max_{\substack{\vartheta' \in \mathcal{N}_{\varepsilon}: \\ d(\vartheta, \vartheta') \ge \varepsilon t}} p(X^{(n)}, \vartheta') \ge p(X^{(n)}, \tilde{\vartheta})\Big),$$

where $\tilde{\vartheta}$ is an element of $\mathcal{N}_{\varepsilon}$ such that $d(\vartheta, \tilde{\vartheta}) \le \varepsilon$. Thus,

$$P_{\vartheta}^{(n)}(d(\vartheta_n^*, \vartheta) \ge \varepsilon t) \le$$

$$\le (\text{card } \mathcal{N}_\varepsilon) \sup_{\substack{\tilde\vartheta,\vartheta' : d(\tilde\vartheta,\vartheta)\le\varepsilon,\\ d(\vartheta,\vartheta')\ge\varepsilon t}} P_\vartheta^{(n)}\left(p(X^{(n)},\vartheta') > p(X^{(n)},\tilde\vartheta) \right) \le$$

$$\le (\text{card } \mathcal{N}_\varepsilon) \exp(-c_d n\varepsilon^\alpha t^\alpha) = \exp(\mathcal{H}(\Theta,d,\varepsilon) - c_d n\varepsilon^\alpha t^\alpha), \quad t\ge t_0.$$

Hence, for $\varepsilon=\varepsilon^*$ we have

$$P_\vartheta^{(n)}(d(\vartheta_n^*,\vartheta) \ge \varepsilon^* t) \le \exp(-c_d n(\varepsilon^*)^\alpha t^\alpha/2), \quad t\ge\max(t_0,t_1).$$

This gives

$$E_\vartheta^{(n)}\left((\varepsilon^*)^{-q} d^q(\vartheta_n^*,\vartheta)\right) = \int_0^{D'} P_\vartheta^{(n)}\left((\varepsilon^*)^{-q} d^q(\vartheta_n^*,\vartheta)\ge t\right)dt =$$

$$= \int_0^{D'} P_\vartheta^{(n)}\left(d(\vartheta_n^*,\vartheta)\ge\varepsilon^* t^{1/q}\right)dt \le$$

$$\le \max(t_0,t_1) + \int_{\max(t_0,t_1)}^{D'} \exp(-c_d n(\varepsilon^*)^\alpha t^{\alpha/q}/2)\,dt \le$$

$$\le \max(t_0,t_1) + D'\exp\left(-\frac{1}{2}c_d n(\varepsilon^*)^\alpha\right)$$

where $D'=(D/\varepsilon^*)^q$. □

5.6. OPTIMAL EDGE ESTIMATORS FOR DUDLEY'S CLASSES

The argument of Section 5.3 can be applied with necessary modifications to general domains in N dimensions, N>2. This will give the edge estimators that have minimax rates of convergence in Hausdorff metric d_∞. Here we consider another approach to high-dimensional edge estimation due to Mammen and Tsybakov (1992). It is based on maximum likelihood on ε-nets. We deal with the d_1-metric. Using Lemma 5.5.3 we show that the MLE on ε-net achieves minimax rates of convergence in d_1-metric on Dudley's classes of domains.

As follows from Dudley (1974) the set $\tilde{\mathcal{G}}$ is compact in d_1-metric and there exists $C_{\mathcal{H}} > 0$ such that

$$(5.18) \qquad \mathcal{H}(\tilde{\mathcal{G}}, d_1, \varepsilon) \le C_{\mathcal{H}}\, \varepsilon^{-(N-1)/\gamma}.$$

In fact, (5.18) is a simple consequence of the well-known entropy bound (Kolmogorov and Tikhomirov (1961)) :

$$(5.19) \qquad \mathcal{H}(\Sigma_{N-1}(\gamma, L), d_{sup}, \varepsilon) = O(\varepsilon^{-(N-1)/\gamma}), \quad \varepsilon \to 0,$$

where d_{sup} is the supremum-norm distance between the functions on B^{N-1}.

Assume that the observations follow the multiplicative model (5.1) and the design satisfy (B1), i.e. X_i's are i.i.d. uniformly distributed in $[0,1]^N$. Let $\mathcal{N}_\varepsilon$ be a minimal ε-net on $\Theta = \tilde{\mathcal{G}}$ in d_1-metric. Define G_n^* as the maximum likelihood estimator on ε-net $\mathcal{N}_\varepsilon$:

$$(5.20) \qquad G_n^* = \underset{G \in \mathcal{N}_\varepsilon}{\operatorname{argmax}}\, p(X^{(n)}, G)$$

where

$$p(X^{(n)}, G) = \prod_{i=1}^{n} p(X_i, Y_i, G),$$

$$p(X_i, Y_i, G) = \begin{cases} 2\bar{p}_0 & \text{if } Y_i = 1,\ X_i \in G \text{ or } Y_i = -1,\ X_i \in \bar{G}, \\ 2(1-\bar{p}_0) & \text{if } Y_i = -1,\ X_i \in G \text{ or } Y_i = 1,\ X_i \in \bar{G}. \end{cases}$$

Here $X^{(n)} = ((X_1, Y_1), \ldots, (X_n, Y_n))$, and $p(X^{(n)}, G)$ is the density of the distribution of $X^{(n)}$ with respect to dominating measure $\nu = (\nu_0 x \nu_1)^n$ where ν_0 is the Lebesgue measure on $[0,1]^N$ and ν_1 is the measure concentrated at points 1, -1, with $\nu_1\{1\} = \nu_1\{-1\} = 1/2$.

One can easily verify that the following definition is equivalent to (5.20) :

$$(5.21) \qquad G_n^* = \underset{G \in \mathcal{N}_\varepsilon}{\operatorname{argmin}} \sum_{i=1}^{n} \Big[I\{X_i \in G\} I\{Y_i = -1\} + I\{X_i \in \bar{G}\} I\{Y_i = 1\} \Big].$$

This shows that G_n^* does not in fact depend on $\bar{p}_0$.

The next theorem gives the convergence rate of G_n^* under the appropriate choice of ε.

5.6.1. THEOREM. Let the observations Y_i be of the form (5.1) and let the design satisfy (B1). Assume that $\varepsilon^* = cn^{-\gamma/(\gamma+N-1)}$ where $c > \left(24C_{\mathcal{H}}/(2\bar{p}_0-1)\right)^{\gamma/(\gamma+N-1)}$. If G_n^* is the maximum likelihood estimator on ε-net (5.21) with $\varepsilon=\varepsilon^*$ then

$$\sup_{G\in\tilde{\mathcal{G}}} E_G(\psi_n^{-q} d_1^q(G,G_n^*)) = O(1),\ n\to\infty,\ \forall q>0,$$

where $\psi_n = n^{-\gamma/(\gamma+N-1)}$.

Proof. Use Lemma 5.5.3. First, we have to check (5.15). Consider the sets $\vartheta=G$, $\vartheta'=G'$, and $\tilde{\vartheta}=\tilde{G}$, such that $d_1(G,\tilde{G})\le\varepsilon$, $d_1(G,G')\ge\varepsilon t$, $t\ge t_0$, for some $t_0>1$. Then, by definition of $p(X^{(n)},G)$,

$$P_{\vartheta}^{(n)}\Big(p(X^{(n)},\vartheta') > p(X^{(n)},\tilde{\vartheta}) \Big) = P_G\Big(\bar{p}_0^{S}(1-\bar{p}_0)^{-S} \ge 1 \Big) = P_G(S\ge 0),$$

where

$$S = \sum_{i=1}^{n}\Big[I\{X_i\in G', Y_i=1\} + I\{X_i\notin G', Y_i=-1\}\Big] -$$

$$- \sum_{i=1}^{n}\Big[I\{X_i\in\tilde{G}, Y_i=1\} + I\{X_i\notin\tilde{G}, Y_i=-1\}\Big].$$

Note that

$$E_G(S) = (2\bar{p}_0-1)[\operatorname{mes}(G\cap(G'\setminus\tilde{G})) - \operatorname{mes}(\bar{G}\cap(G'\setminus\tilde{G})) -$$

$$- \operatorname{mes}(G\cap(\tilde{G}\setminus G')) + \operatorname{mes}(\bar{G}\cap(\tilde{G}\setminus G'))] =$$

$$= (2\bar{p}_0-1)[d_1(G,\tilde{G}) - d_1(G,G')] \le (2\bar{p}_0-1)[\varepsilon - d_1(G,G')].$$

Note that $S = \sum_{i=1}^{n}\eta_i$, where η_i are independent random variables, $|\eta_i|\le 1$, and $\operatorname{Var}(\eta_i)\le E(\eta_i^2) = d_1(\tilde{G},G')\le d_1(G,G')+d_1(G,\tilde{G}) \le d_1(G,G')+\varepsilon$. Applying the Bernstein inequality and using these estimations we find

$$P_G(S\ge 0) \le P_G\left(\frac{1}{n}\sum_{i=1}^{n}(\eta_i - E(\eta_i)) \ge (2\bar{p}_0-1)[\varepsilon - d_1(G,G')]\right) \le$$

$$\le \exp\left(- \frac{n[(2\bar{p}_0-1)(d_1(G,G')-\varepsilon)]^2}{4(d_1(G,G')+\varepsilon)} \right).$$

Put $t_0=2$. Then for $d_1(G,G') \ge t\varepsilon$, $t\ge t_0$, we have

$$P_G(S\ge 0) \le \exp\left(- [(2\bar{p}_0-1)^2/12]n\varepsilon t \right),$$

which proves (5.15) with $c_d=(2\bar{p}_0-1)^2/12$, $\alpha=1$, $t_0=2$.

To prove (5.16) with $\alpha=1$ note that by (5.18) and by definition of ε^*

$$\text{(5.22)} \qquad \mathcal{H}(\tilde{\mathcal{G}},d_1,\varepsilon^*) \le C_{\mathcal{H}} (\varepsilon^*)^{-(N-1)/\gamma} \le \tfrac{1}{2}c_d n\varepsilon^* t, \quad t\ge 1.$$

Hence we can apply Lemma 5.5.3. Since in our case D=1, $\alpha=1$, $t_0=2$, $t_1=1$, we get

$$\sup_{G\in\tilde{\mathcal{G}}} E_G\left((\varepsilon^*)^{-q}d_1^q(G_n^*,G)\right) \le 2 + (1/\varepsilon^*)^q\exp\left(-\tfrac{1}{2}c_d n\varepsilon^*\right).$$

This inequality and the definition of ε^* entail the theorem. □

5.6.2. REMARK. The assumptions of multiplicative model and the design (B2) in Theorem 5.6.1 can be weakened. It can be extended to the design (B2) and to the model with additive noise (3.1), provided the distribution of ξ_i's satisfies Assumption 5.1.5. Also the extension to arbitrary power loss functions is obvious in view of Lemma 5.5.3. Proof of these extensions is left to the reader.

Consider now the edge estimation problem for convex domains. Here we show that in this problem, again, one can use MLE's on ε-nets to get upper bounds on minimax risks. Lower bounds are obtained as a simple modification of Theorem 5.1.1 (see Mammen and Tsybakov (1992) for details).

Denote by $\mathcal{G}_{conv}$ the set of all closed convex subsets of $K=[0,1]^N$. Dudley (1974) proved that there exists $C_{conv}>0$ such that

$$(5.23) \qquad \mathcal{H}(\mathcal{G}_{conv}, d_1, \varepsilon) \le C_{conv}\, \varepsilon^{-(N-1)/2},$$

This inequality shows that the class of all convex domains has roughly the same complexity as the Dudley's class $\tilde{\mathcal{G}}$ with smoothness parameter $\gamma=2$.

5.6.3. THEOREM. Let the observations Y_i be of the form (5.1) and let the design satisfy (B1). Assume that $\varepsilon^*=cn^{-2/(N+1)}$ where $c > \left(24C_{conv}/(2\bar{p}_0-1)\right)^{2/(N+1)}$. If G_n^* is the maximum likelihood estimator (5.21) where $\mathcal{N}_\varepsilon$ is the minimal ε-net on $\mathcal{G}_{conv}$ in d_1-metric, with $\varepsilon=\varepsilon^*$, then

$$(5.24) \qquad \sup_{G\in\mathcal{G}_{conv}} E_G(\psi_n^{-q} d_1^q(G,G_n^*)) = O(1), \; n\to\infty,$$

with $\psi_n=n^{-2/(N+1)}$. Moreover, ψ_n is the A-minimax rate of convergence for the class $\mathcal{G}_{conv}$.

<u>Proof</u> of (5.24) differs from the proof of Theorem 5.6.1 only in the bound on ε-entropy. Instead of (5.18) one has to apply (5.23). For A-minimax lower bounds we refer to Mammen and Tsybakov (1992).

5.7. ON CALCULATION OF OPTIMAL EDGE ESTIMATORS FOR GENERAL DOMAINS

Theorem 5.6.1 is mainly the existence result. Together with Theorem 5.1.1 it shows that the minimax rate of convergence for Dudley's classes of domains in dimensions $N\ge 2$ is of order $n^{-\gamma/(\gamma+N-1)}$. The problem of constructing practical estimators which achieve this rate for such domains remains open. Although, there exist special cases for which the problem is solved, and some conjectures that we would like to mention here.

First, for sets with Lipschitz boundaries ($\gamma=1$) the optimal estimators are based on the simple classification procedure of

Section 5.2 (although, the d_1-optimality is achieved up to the log-factor).

For $\gamma>1$ the calculation of optimal estimators is a hard problem, even if one deals with the boundary fragments.

Now we discuss an important class of sets where the computational problem is reduced to that of boundary fragments.

Let N=2. The domain G is called <u>star-shaped</u> if its center

$$x^* = (\text{mes } G)^{-1}\int_G x\, dx$$

is its inner point, and the boundary $\Gamma=\partial G$ is defined as

$$\Gamma = \left\{x\in[0,1]^2:\ x=x^*+\Delta,\ \Delta=(g(\phi)\cos\phi,\ g(\phi)\sin\phi),\ 0\le\phi<2\pi\right\}$$

where $g(\phi)>0$ is a function that belongs to the smoothness class $\Sigma(\gamma,L)$. In polar coordinate system centered at x^* one can represent Γ as a curve where r is the radius and ϕ is the polar angle. In this coordinate system we define the new data $X_i' = (r_i,\phi_i)$ where ϕ_i is the polar angle corresponding to X_i and $r_i=g(\phi_i)$. Clearly, if $G\subseteq[0,1]^2$ then $0\le r_i<\sqrt{2}$, $0\le\phi_i<2\pi$. Thus, for x^* known, the estimation of G is reduced to edge estimation for boundary fragments on $K=[0,\sqrt{2}]\times[0,2\pi]$, and the optimal procedure of Chapter 4 can be applied with minor changes. The estimation of star-shaped sets with known center x^* was studied by Rudemo, Skovgaard and Stryhn (1990), Rudemo and Stryhn (1991) who proved consistency of the piecewise-constant estimator ($\gamma=1$) and pointed out some interesting applications.

If the center x^* is unknown the initial step is to be added for calculation of its estimator $\hat{x}^*$. This could be, for example,

$$\hat{x}^* = (\text{mes } G_n)^{-1}\int_{G_n} x\, dx$$

where G_n is some preliminary estimator of the domain G. However, the only calculable estimator G_n for <u>general</u> domains that we are

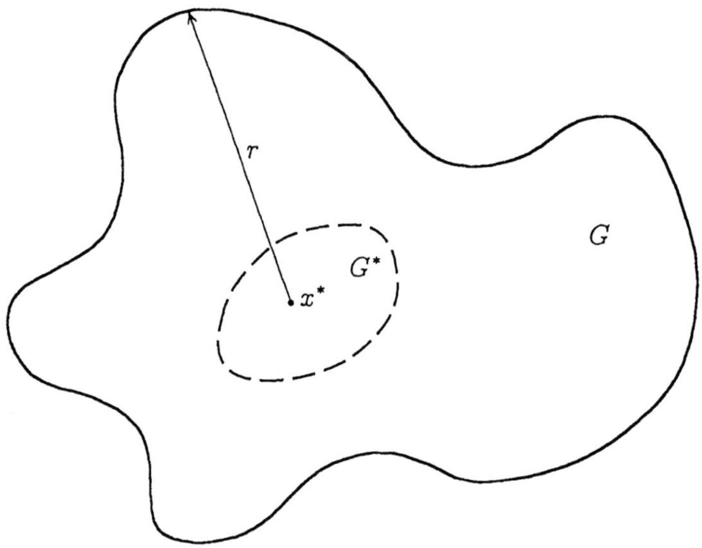

Solid line - the boundary of the domain G.
Dotted line - the boundary of the inner set G^*.
The domain G is star-shaped around any point $x^* \in G^*$.

Fig. 5.6.

aware of is $G_n = G_n^{**}$ introduced in Section 5.2. We know that G_n^{**} converges rather slowly: its rate is optimal only for the case of $\gamma=1$. Hence, with $G_n = G_n^{**}$ the estimator $\hat{x}^*$ is too rough. This makes the final edge estimator also rough. To avoid this effect let us make a reasonable change in the definition of star-shaped set. Assume that there exists an open subset G^* of G such that x^* is the inner point of G^*, and for any $x_0 \in G^*$ the boundary Γ can be represented as

$$\Gamma = \left\{ x\in[0,1]^2 :\ x=x_0+\Delta,\ \Delta=(g(x_0,\phi)\ \cos\phi,\ g(x_0,\phi)\ \sin\phi),\ 0\le\phi<2\pi \right\}$$

where $g(x_0,\phi)$ belongs to the class $\Sigma(\gamma,L)$ for all $x_0 \in G^*$. Now, if G_n is just a consistent estimator of G in d_1-metric then for n large enough $\hat{x}^* \in G^*$ with probability close to 1. Hence the problem can be reduced to boundary fragments as described earlier, with $\hat{x}^*$ as a central point.

CHAPTER 6. IMAGE RECONSTRUCTION UNDER RESTRICTIONS ON ESTIMATES

The results of Chapters 3-5 show that the best possible rates of convergence in various image and edge estimation problems are attained by estimators that are complicated from the computational viewpoint. On the other hand, there are many data processing algorithms in image analysis which deal with some simple partial procedures and do apply in practice. We study here two popular and important classes of such procedures: linewise processing and linear estimates. The questions are: whether they ensure the minimax rate of convergence and, if not, what is the best possible rate within these classes?

6.1. NAÏVE LINEWISE PROCESSING

The study of linewise procedures reveals in the most obvious form the link between image estimation, change-point problem, and nonparametric regression. Consider the simplest two-dimensional case of edge estimation for boundary fragments. In this case the true domain G, $G \in \mathcal{G}_0(\gamma, L, h)$, is given by

$$G = \left\{ (x_1, x_2):\ 0 \le x_1 \le 1,\ 0 \le g(x_1) \le x_2 \right\},\quad g(x_1) \in \Sigma(\gamma, L, h),\quad \gamma > 1 .$$

For technical reasons it is convenient to consider the design which has not yet been studied in this book but which is quite close to the designs (A2) and (B1) defined in Section 3.2.1. Let $n_1 = n^{1/2}$ be an integer, and let the design points be

$$(6.1)\qquad X_{ij} = (i/n_1,\ j/n_1 - \eta_i),\quad i,j = 1,\dots,n_1,$$

where η_i are independent random variables uniformly distributed in $[0, n_1^{-1}]$. Design (6.1) is composed of random shifts of the regular grid which are independent for each line $x_1 = i/n_1$, $i = 1,\dots,n_1$. Consider the observations

$$(6.2)\qquad Y_{ij} = I(X_{ij} \in G) + \xi_{ij},\quad i,j = 1,\dots,n_1,$$

where ξ_{ij} are i.i.d. $(0,\sigma^2)$-Gaussian random variables.

Though the problem of edge estimation from observations (6.2) was not formally covered by the results of the previous chapters due to the particular nature of the design (6.1), nevertheless one can easily verify that the minimax rate of convergence is equal to $\psi_n = n^{-\gamma/(\gamma+1)}$ for $\gamma > 1$. An optimal edge estimator may be presented, e.g. by the MLE defined in Section 4.1.

Note that for each line $x_1 = i/n_1$ we have now an independent change-point problem with the change-point parameter $\vartheta_i = g(i/n_1)$, $i=1,\dots,n_1$. Denote by ϑ^*_{in} the maximum likelihood estimator (MLE) for ϑ_i obtained from observations

$$((X_{i1},Y_{i1}),\dots,(X_{in_1},Y_{in_1})), \quad i=1,\dots,n_1.$$

Here the subscript "n" in ϑ^*_{in} indicates the total number of observations that tends to infinity in our setup. The MLE's ϑ^*_{in} are independent for different i and their properties are discussed in detail in Section 1.9.

Corollary 1.9.6 states that the stochastic expansions

$$(6.3) \qquad \vartheta^*_{in} = g(i/n_1) + b_{in_1} + n_1^{-1}\varepsilon_{in_1}, \quad i=1,\dots,n_1,$$

are valid for all n with zero-mean random errors ε_{in_1} which are independent for different i and have bounded variances. The bias terms b_{in_1} depend on $g(x_1)$ and for all n large enough they satisfy the inequalities

$$(6.4) \qquad |b_{in_1}| \le \exp(-\lambda n_1), \quad i=1,\dots,n_1,$$

with some positive λ uniformly over $g(x_1) \in \Sigma(\gamma, L, h)$.

We mean by naïve linewise processing the following two-step procedure. The first step consists in calculation of ϑ^*_{in} for all the lines $x_1 = i/n_1$. At the second step some smoothing algorithm is applied to ϑ^*_{in}'s which are regarded now as new observations in nonparametric regression

$$\vartheta_{in}^{*} \approx g(i/n_1) + \varepsilon_{in_1}/n_1 \ , \ i=1,\dots,n_1. \tag{6.5}$$

Surely, the bias terms b_{in_1} in (6.3) are negligible as follows from (6.4). The error of edge estimation in d_1-metric equals to the L_1-error of estimation of regression function g from observations (6.5). Take for definiteness the piecewise-polynomial regression estimator $\hat{g}_n(x_1)$, $0 \le x_1 \le 1$, corresponding to the bin width δ_n, $\delta_n \to 0$, $n_1\delta_n \to \infty$. The bias and variance terms related to this estimator are equal to $O(\delta_n^{\gamma})$ and $O\left((1/n_1)^2(n_1\delta_n)^{-1}\right)$ respectively as $n \to \infty$.

To show this apply Theorem 1.6.1 and note that the noise in (6.5) includes the vanishing factor $1/n_1$ while the total number of observations is $n_1\delta_n$. Balancing the squared bias and the variance (cf. Section 1.6), we get $\delta_n^{2\gamma} = n_1^{-2}(n_1\delta_n)^{-1}$. Hence, the optimal bin width is $\delta_n^{*} = n^{-3/(4\gamma+2)}$ and the rate of convergence for the piecewise-polynomial estimator with $\delta_n = \delta_n^{*}$ is given by

$$E_G\left[\int_0^1 |\hat{g}_n(x_1) - g(x_1)|dx_1\right] = O(n^{-3\gamma/(4\gamma+2)}) \text{ as } n \to \infty. \tag{6.6}$$

Two conclusions are appropriate here. First, the rate of convergence obtained by the linewise processing is not the minimax one. Indeed, inequality $\gamma/(\gamma+1) > 3\gamma/(4\gamma+2)$ holds for $\gamma>1$. Second, the rate of convergence achieved in (6.6) cannot be improved uniformly in $g \in \Sigma(\gamma, L, h)$ if we use only the MLE ϑ_{in}^{*} at the second step of the procedure. We omit the proof of this statement since it follows directly the idea of Example 2.7.3.

Now we concentrate on the next problem: whether it is possible to modify the naïve linewise processing in such a way that the resulting procedure would achieve the minimax rate of convergence? The answer is positive. Certainly, for such a procedure we must have not only the values ϑ_{in}^{*} but also some additional information for each line $x_1 = i/n_1$.

6.2. MODIFIED LINEWISE PROCEDURE

The idea of modification is rather simple. Suppose that at the point $X_i^* = (i/n_1, \vartheta_{in}^*)$ we are able to get a new observation $Y_i^* = I(X_i^* \in G) + \zeta_i$ where ζ_i are i.i.d. zero-mean random variables. Then we can try the least squares edge estimator (see Section 4.1) based on the new observations (X_i^*, Y_i^*), $i=1,\dots,n_1$. Since the new design points X_i^* are very close to the edge one can expect that such estimator will improve upon the naïve procedure. Now we show that for each i it is possible to find an observation $Y_i^* \in \{Y_{ij},\ j=1,\dots,n_1\}$ which has the properties that we supposed.

In addition to the assumptions of the previous section let us suppose that $n_1 = n^{1/2}$ is even. Again, as in Section 4.4, split each sample $(\ (X_{i1}, Y_{i1}),\dots,(X_{i,n_1}, Y_{i,n_1})\ )$, $i=1,\dots,n_1$, into odd and even subsamples. Denote by ϑ_{in}^{odd} and ϑ_{in}^{even} the maximum likelihood estimators obtained respectively from the odd and even subsamples, i. e. from

$$(\ (X_{i1}, Y_{i1}),\ (X_{i3}, Y_{i3}),\dots,(X_{i,n_1-1}, Y_{i,n_1-1})\ )$$

and

$$(\ (X_{i2}, Y_{i2}),\ (X_{i4}, Y_{i4}),\dots\ ,(X_{i,n_1}, Y_{i,n_1})\ ).$$

By definition the MLE ϑ_{in}^{odd} is the average:

$$(i/n_1, \vartheta_{in}^{odd}) = (X_{i,\hat{j}-1} + X_{i,\hat{j}+1})/2 \tag{6.7}$$

for some $\hat{j}$, $1 < \hat{j} < n_1$, $\hat{j}$ even (see Section 1.9). For the random shift of regular design the right-hand side of (6.7) equals $X_{i,\hat{j}}$. This is the only point where the technical convenience of the design (6.1) appears: the MLE strictly coincides with one of design points. Identity (6.7) holds for ϑ_{in}^{even} as well but now $\hat{j}$ is an odd integer.

Let the even integer j_0 be such that

$$(i/n_1, \vartheta_i) \in [X_{i,j_0-1}, X_{i,j_0+1})$$

where $[X_{i,j_0-1}, X_{i,j_0+1})$ denotes the segment of the vertical straight line that connects X_{i,j_0-1} and X_{i,j_0+1}. Define $\bar{\vartheta}_i^{odd}$ as the second coordinate of the midpoint of this segment, i.e.

$$(i/n_1, \bar{\vartheta}_i^{odd}) = (X_{i,j_0-1} + X_{i,j_0+1})/2 = X_{i,j_0}, \quad i=1,\dots,n_1.$$

The definition of $\bar{\vartheta}_i^{even}$ is quite similar, with j_0 odd.

Now we want to find the distribution functions of the random variables $n_1(\vartheta_{in}^{odd}-\vartheta_i)$ and $n_1(\vartheta_{in}^{even}-\vartheta_i)$. Following the lines of Proposition 1.9.3 (under the design (A2)), one can easy show that these random variables are bounded in P_G-probability uniformly over $\vartheta_i \in (h, 1-h)$, $i=1,\dots,n_1$, i. e. uniformly over $g(x_1) \in \Sigma(\gamma, L, h)$. But unlike the case of Proposition 1.9.4, these random variables are biased:

$$\liminf_{n\to\infty} \sup_{G \in \mathcal{G}_0(\gamma,L,h)} E_G(n_1(\vartheta_{in}^{odd}-\vartheta_i)) > 0.$$

The reason is quite obvious: the distribution of ϑ_{in}^{odd} doesn't change if ϑ_i varies within the interval $((j_0-1)/n_1, j_0/n_1)$, j_0 even, since the corresponding points $(i/n_1, \vartheta_i)$ are covered by the same design interval $[X_{i,j_0-1}, X_{i,j_0+1})$ under any "shift value" η_i, $0<\eta_i<n_1^{-1}$. The same is true for ϑ_{in}^{even}: the random shift covers only one half part of each design interval. As in Lemma 1.9.5, we first find the distribution of the difference $n_1(\bar{\vartheta}_i^{odd}-\vartheta_i)$. Unfortunately, this difference is not uniformly distributed as was the case in Lemma 1.9.5. Routine considerations show that the random variable $n_1(\bar{\vartheta}_i^{odd}-\vartheta_i)$ is uniformly distributed in

$$(6.8) \qquad \begin{cases} (-1,-\tau_i) \cup (1-\tau_i,1) & \text{if } [n_1\vartheta_i] \text{ is even} \\ (-\tau_i, 1-\tau_i) & \text{if } [n_1\vartheta_i] \text{ is odd} \end{cases}$$

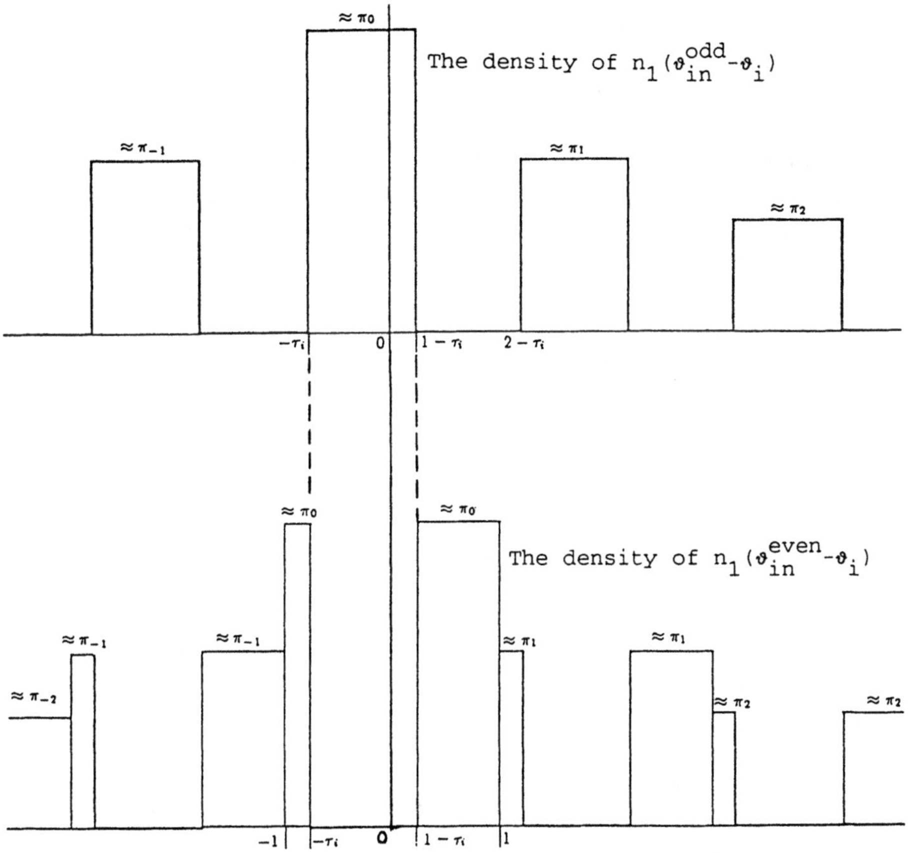

Fig. 6.1. The saw-tooth densities approximating the distributions of MLE's from subsamples

where $\tau_i = \tau_i(\vartheta_i) = n_1\vartheta_i - [n_1\vartheta_i]$, $0 \le \tau_i < 1$. Similarly, the random variable $n_1(\bar{\vartheta}_i^{even} - \vartheta_i)$ is uniformly distributed in

$$(6.9) \qquad \begin{cases} (-1, -\tau_i) \cup (1-\tau_i, 1) & \text{if } [n_1\vartheta_i] \text{ is odd} \\ (-\tau_i, 1-\tau_i) & \text{if } [n_1\vartheta_i] \text{ is even.} \end{cases}$$

Hence, as in Lemma 1.9.5, one may expect that the distribution functions of $n_1(\vartheta_{in}^{odd} - \vartheta_i)$ and $n_1(\vartheta_{in}^{even} - \vartheta_i)$ are approximated by the distributions having piecewise constant density functions (see Fig. 6.1). But, unlike the density function $\pi(x)$ in Lemma 1.9.5, they have a saw-tooth shape with periodical zero gaps. The period of zero gaps is equal to 1, and the gaps for $n_1(\vartheta_{in}^{odd} - \vartheta_i)$ alternate with those for $n_1(\vartheta_{in}^{even} - \vartheta_i)$. Thus, it is reasonable to mix the estimators ϑ_{in}^{odd} and ϑ_{in}^{even} randomly, i.e. to define

$$(6.10) \qquad \vartheta_{in}^{*} = \begin{cases} \vartheta_{in}^{odd} & \text{with probability } 1/2 \\ \vartheta_{in}^{even} & \text{with probability } 1/2 \end{cases}$$

where some independent stochastic procedure is applied to make randomization in (6.10). In Remark 6.2.4. below we show that it is possible to define a non-randomized estimator equivalent to (6.10).

6.2.1. LEMMA. The density of the random variable $n_1(\vartheta_{in}^{*} - \vartheta_i)/2$ satisfies the statement of Lemma 1.9.5 uniformly in $g(x_1) \in \Sigma(\gamma, L, h)$.

Proof is the same as in Lemma 1.9.5: the discrete distributions

$$P_G\left(n_1(\vartheta_{in}^{odd} - \bar{\vartheta}_i^{odd})/2 = k \mid X_1, \ldots, X_n\right) \approx \pi_k$$

and

$$P_G\left(n_1(\vartheta_{in}^{even} - \bar{\vartheta}_i^{even})/2 = k \mid X_1, \ldots, X_n\right) \approx \pi_k$$

must be convolved with (6.8) and (6.9) and then mixed randomly. □

Define $Y_i^* = Y_{i,j^*}$ where Y_{i,j^*} are observations (6.2) at the design points $X_i^* = (i/n_1, \vartheta_{in}^*) = X_{i,j^*}$, $i=1,\ldots,n_1$ (here j^* is randomized as well as ϑ_{in}^*). Note that

$$(6.11) \qquad Y_i^* = I(X_i^* \in G) + \zeta_i \ , \quad i=1,\ldots,n_1,$$

where $\zeta_i = \xi_{i,j^*}$ are i.i.d. $(0,\sigma^2)$-Gaussian random variables; ζ_i is independent of ϑ_{in}^*.

At the first step of the modified procedure the pairs $(Y_i^*, \vartheta_{in}^*)$ are calculated. We will show that the total necessary additional information to ϑ_{in}^* is concentrated in Y_i^* : the knowledge of pairs $(Y_i^*, \vartheta_{in}^*)$ is sufficient to get the minimax linewise procedure.

In the following we use the notation of Sections 4.1-4.3 properly fitted to the design (6.1). Thus, let $\delta_n = n^{-1/(\gamma+1)}$, $M = M_n = 1/\delta_n$. The slices A_ℓ and domains $D_\ell = G \cap A_\ell$, $B_\ell(\theta)$, $\Theta^{(K)}$, Θ_n are defined as in Section 4.1.

The second step of the modified linewise procedure is to construct the least squares estimator (LSE) similar to the MLE in Section 4.1: let $\hat{\theta}_\ell = (\hat{\theta}_{0\ell}, \ldots, \hat{\theta}_{k\ell})$ be a solution of

$$(6.12) \qquad J_\ell^*(\hat{\theta}_\ell) = \max_{\theta \in \Theta_n} J_\ell(\theta)$$

where

$$J_\ell^*(\theta) = \sum_{i=1}^{n_1} I\left(X_i^* \in B_\ell(\theta)\right) Y_i^* - (1/2) \operatorname{card}\left\{i : X_i^* \in B_\ell(\theta)\right\}.$$

Consider the edge estimator

$$\hat{g}_\ell(x_1) = \hat{\theta}_{0\ell} + \hat{\theta}_{1\ell}(x_1 - u_{\ell-1}) + \ldots + \hat{\theta}_{k\ell}(x_1 - u_{\ell-1})^k,$$

$$u_{\ell-1} \le x_1 < u_\ell, \ \ell = 1, \ldots, M.$$

The domain estimator G_n^* is then the closure of the set

$$\left\{(x_1,x_2):\ 0\le x_2\le \hat{g}_\ell(x_1)\right\}.$$

The estimator $\hat{\theta}_\ell=(\hat{\theta}_{0\ell},\ldots,\hat{\theta}_{k\ell})$ is not the MLE since it does not account for the dependence of the distribution of ϑ^*_{in} on an unknown parameter θ. Nevertheless, the following theorem states that it leads to edge estimators having the best possible rates of convergence.

6.2.2. THEOREM. Let, for each $i=1,\ldots,n_1$, ϑ^*_{in} be the estimators (6.10) of the change-point parameters $\vartheta_i=g(i/\sqrt{n})$, based on the observations (6.2). Then the domain estimator G^*_n has the I-minimax rate of convergence $\psi_n=n^{-\gamma/(\gamma+1)}$ for the class of boundary fragments $\mathcal{G}_0(\gamma,L,h)$ in d_1-metric.

6.2.3. REMARK. The amount of information which is to be added to each ϑ^*_{in}, $i=1,\ldots,n_1$, may be reduced to one bit per line Indeed, if we define $Y'_i=\text{sign}(Y^*_i-1/2)$, then the LSE obtained from (Y'_i,ϑ^*_{in}), $i=1,\ldots,n_1$, guarantees the same rate of convergence as that in Theorem 6.2.2. One can get this result following the argument of Section 4.4.

6.2.4. REMARK. In this book we do not study the randomized estimators generated by some external stochastic influence, neither we need them now. Indeed, consider the estimator ϑ^{odd}_{in} as it was defined above and ϑ^{even}_{in} obtained from all the even observations but the last one. If n is large enough to ensure the inequality $X_{i,n_1}>1-h$, then $Y_{i,n_1}=\xi_{i,n_1}$ is zero-mean Gaussian. Hence, the estimator

$$\vartheta^*_{in}=\begin{cases}\vartheta^{odd}_{in} & \text{if}\quad Y_{i,n_1}<0\\ \vartheta^{even}_{in} & \text{if}\quad Y_{i,n_1}\ge 0\end{cases}$$

satisfies all the demands: it is non-randomized and at the same time its distribution is as good as that in Lemma 1.9.5. Surely, this construction is a pure mathematical trick.

6.3. PROOFS

Here we use the notation of Sections 4.1, 4.2. Assume without loss of generality that $M_n=1/\delta_n$ and $n_1\delta_n=n^\alpha$ are some integers which tend to infinity as $n \to \infty$. Here $\alpha=(\gamma-1)/(2(\gamma+1))$ is positive for $\gamma>1$. Introduce the next notation for the part of the line $x_1=i/n_1$ which belongs to the domain $B_\ell(\theta)\Delta D_\ell$:

$$\mathcal{L}_i = \mathcal{L}_i(\theta) = \Big\{(x_1,x_2):\ x_1=i/n_1\Big\} \cap \Big(B_\ell(\theta)\ \Delta\ D_\ell\Big)$$

By definition, $\mathcal{L}_i$ are connected intervals. Let $|\mathcal{L}_i|$ be the length of the interval $\mathcal{L}_i$, and let

$$I_\ell = \Big\{i:\ u_{\ell-1} \le i/n_1 < u_\ell\Big\}$$

be the set of indices i such that $\mathcal{L}_i \subset A_\ell$. Note that

$$\text{card } I_\ell = n_1\delta_n = n^\alpha \to \infty \quad \text{as} \quad n \to \infty.$$

As in Section 4.2 we associate with θ a set of integers $m=(m_0,\dots,m_k)$ and mark this correspondence by $\theta=\theta(m)$. Remind that the set of integers $m_\ell^{(0)}$ corresponds to the Taylor approximation of the edge function $g(x_1)$ in the interval $[u_{\ell-1},u_\ell)$ and $\hat{m}_\ell$ corresponds to the estimator $\hat{\theta}_\ell$ in (6.12): $\hat{\theta}_\ell=\theta(\hat{m}_\ell)$.

The problem (6.12) is equivalent to

$$\max_{\theta\in\Theta_n} Z_\ell^*(\theta), \qquad Z_\ell^*(\theta) = J_\ell^*(\theta) - J_\ell^{(G)}, \tag{6.13}$$

where

$$J_\ell^{(G)} = \sum_{i\in I_\ell} I\Big(X_i^*\in G\Big)\ Y_i^* - (1/2)\ \text{card}\Big\{i: X_i^*\in G\Big\}$$

is independent of θ.

6.3.1. LEMMA. For each ℓ, $\ell=1,\dots,M$, the random function $Z_\ell^*(\theta)$ can be presented in the form

$$Z_\ell^*(\theta) = \Delta_\ell^*(\theta) - (1/2) \sum_{i\in I_\ell} P_G\Big(X_i^*\in\mathcal{L}_i(\theta)\Big)$$

where the random variables $\Delta_\ell^*(\theta)$ are independent for different ℓ,

$E_G\left(\Delta_\ell^*(\theta)\right) = 0$ and for all z small enough

$$E_G\left(\exp(z\Delta_\ell^*(\theta))\right) \le \exp(\varkappa_0 z^2 \sum_{i\in I_\ell} P_G\left(X_i^* \in \mathcal{L}_i(\theta)\right)) \tag{6.14}$$

with a positive $\varkappa_0$ which does not depend on n, ℓ, θ and G.

Proof. Note that (6.11) entails the representation

$$Z_\ell^*(\theta) = \sum_i I\left(X_i^* \in B_\ell(\theta)\setminus D_\ell\right) \zeta_i - \sum_i I\left(X_i^* \in D_\ell\setminus B_\ell(\theta)\right) \zeta_i - (1/2)\ \Lambda_\ell^*$$

where $\Lambda_\ell^* = \text{card}\left\{i:\ X_i^* \in B_\ell(\theta)\Delta D_\ell\right\}$ with the mean value

$$E_G(\Lambda_\ell^*) = \sum_{i\in I_\ell} P_G\left(X_i^* \in \mathcal{L}_i(\theta)\right)$$

The rest of proof is similar to that in Lemma 4.2.2, in particular, inequality (6.14) is the direct analogue to (4.4). □

6.3.2. LEMMA. For n large enough and $\theta\in\Theta_n$

$$(1/2)\ d_1\left(B_\ell(\theta), D_\ell\right) \le (1/n_1) \sum_{i\in I_\ell} |\mathcal{L}_i(\theta)| \le 2\ d_1\left(B_\ell(\theta), D_\ell\right)$$

where $n_1 = n^{1/2}$.

Proof. Note that $(1/n_1) \sum_{i\in I_\ell} |\mathcal{L}_i(\theta)|$ is the Riemann sum for

$$\int_{u_{\ell-1}}^{u_\ell} |\theta_0+\theta_1(x_1-u_{\ell-1})+\ldots+\theta_k(x_1-u_{\ell-1})^k - g(x_1)|dx_1 = d_1\left(B_\ell(\theta), D_\ell\right).$$

The latter quantity is approximated by the Riemann sum with accuracy

$$O\left(d_1\left(B_\ell(\theta), D_\ell\right) \ /\ \text{card}\ I_\ell\right) \le (1/2)\ d_1\left(B_\ell(\theta), D_\ell\right)$$

for all sufficiently large n. □

6.3.3. LEMMA. If $\|m-m_\ell^{(0)}\|\le n^\alpha/(\gamma+1)$ and $\theta=\theta(m)$, then for all $\ell=1,\dots,M$, and n large enough the following inequalities hold:

$$\max_{i\in I_\ell} |\mathcal{L}_i(\theta)| \le 1/n_1 \; ; \tag{6.15}$$

$$n_1\Big(c_1|\mathcal{L}_i(\theta)|-\exp(-\lambda_0 n_1)\Big) \le P_G\Big(X_i^*\in\mathcal{L}_i(\theta)\Big) \le \tag{6.16}$$
$$\le n_1\Big(c_1|\mathcal{L}_i(\theta)|-\exp(-\lambda_0 n_1)\Big);$$

$$(1/2)c_1 d_1\Big(B_\ell(\theta),D_\ell\Big)-\exp(-\lambda_0 n_1) \le \sum_{i\in I_\ell} P_G\Big(X_i^*\in \mathcal{L}_i(\theta)\Big) \le \tag{6.17}$$
$$\le 2c_2 d_1\Big(B_\ell(\theta),D_\ell\Big)+\exp(-\lambda_0 n_1)$$

with some positive constants c_1, c_2, and λ_0.

Proof. Similarly to (4.13) and (4.14) we have

$$\max_{i\in I_\ell} |\mathcal{L}_i(\theta)| \le \delta_n^\gamma\, L/k! + \delta_n^\gamma(k+1)\|m-m_\ell^{(0)}\| \le$$
$$\le \delta_n^\gamma\, L/k! + \delta_n^\gamma n^\alpha(k+1)/(\gamma+1) \le$$
$$\le \delta_n^\gamma\, n^\alpha\, (L/k!\; n^{-\alpha} + (k+1)/(\gamma+1)) \le \delta_n^\gamma\, n^\alpha = n^{-1/2} = 1/n_1$$

which proves (6.15). Note that

$$P_G\Big(X_i^*\in\mathcal{L}_i(\theta)\Big) = P_G\Big(n_1(\vartheta_{in}^*-\vartheta_i)/2 \in (n_1/2)(\mathcal{L}_i(\theta)-\vartheta_i)\Big)$$

where $(\mathcal{L}_i(\theta)-\vartheta_i)$ is the shift of the interval $\mathcal{L}_i(\theta)$. By definition, one of the ends of $(\mathcal{L}_i(\theta)-\vartheta_i)$ is the origin. Hence, Lemma 6.2.1 applies which states that the distribution function of $(n_1/2)(\vartheta_{in}^*-\vartheta_i)$ is approximated by that having the density $\pi(x)$. Taking into account (6.15), one gets (6.16). The inequalities (6.17) are the direct consequence of Lemma 6.3.2 and (6.16). □

6.3.4. LEMMA. If $\|m-m_\ell^{(0)}\| > n^\alpha/(\gamma+1)$ and $\theta=\theta(m)\in\Theta_n$, then

$$\sum_{i\in I_\ell} P_G\Big(X_i^*\in \mathcal{L}_i(\theta(m))\Big) \ge c_3 n^\alpha \tag{6.18}$$

for all n large enough, $c_3 > 0$.

Proof. Assume for definiteness that $g(i/n_1) \le g_\ell(i/n_1;\theta(m))$ where $g(x_1)$ is the edge function, and

$$g_\ell(x_1;\theta) = \theta_{0\ell}+\theta_{1\ell}(x_1-u_{\ell-1})+\ldots+\theta_{k\ell}(x_1-u_{\ell-1})^k, \quad u_{\ell-1}\le x_1<u_\ell .$$

Then

$$P_G\left(X_i^* \in \mathcal{L}_i(\theta)\right) = P_G\left(g(i/n_1) \le \vartheta_{in}^* \le g_\ell(i/n_1;\theta(m)) \right)$$

and

$$g_\ell(x_1;\theta(m))-g(x_1) = \left[\sum_{j=0}^{k} \left(\theta_j(m) - \frac{g^{(j)}(u_{\ell-1})}{j!}\right) (x_1-u_{\ell-1})^j\right] +$$

$$+ \left[\sum_{j=0}^{k} \frac{g^{(j)}(u_{\ell-1})}{j!} (x_1-u_{\ell-1})^j - g(x_1)\right], \quad u_{\ell-1} \le x < u_\ell.$$

Since the last summand presents the Taylor approximation, we have

$$g_\ell(x_1;\theta(m))-g(x_1) \ge \delta_n^\gamma \left(\sum_{j=0}^{k} (m_j - m^{(0,j)})y^j - L/k!\right) \ge$$

$$\ge \delta_n^\gamma \, \|m-m_\ell^{(0)}\| \left(\sum_{j=0}^{k} (m_j - m^{(0,j)})\|m-m_\ell^{(0)}\|^{-1}y^j - L/k!\,\|m-m_\ell^{(0)}\|^{-1}\right)$$

where $y=(x_1-u_{\ell-1})/\delta_n$. Since $\|m-m_\ell^{(0)}\| > n^\alpha/(\gamma+1)$ the last term is negligible as n tends to infinity. There exists an interval U of length $|U| > 0$ and a positive c_4 such that

$$\sum_{j=0}^{k} (m_j - m^{(0,j)})\|m-m^{(0)}\|^{-1}y^j \ge c_4 \quad \text{for } y\in U.$$

Indeed, the set of polynomials in the left-hand side is compact and separated from zero. It means that among $\left\{\mathcal{L}_i(\theta(m)),\ i\in I_\ell\right\}$ in (6.18) there exist at least $|U|n_1\delta_n = |U|n^\alpha$ intervals with lengths exceeding

$$\delta_n^\gamma/(\gamma+1) \ge \delta_n^{\gamma+1}n_1/(\gamma+1) = n_1^{-1}/(\gamma+1).$$

For n large each of the corresponding probabilities $P_G\left(X_i^*\in\mathscr{L}_i(\theta(m))\right)$ is bounded from below by a positive constant in accordance with (6.16). This proves (6.18).□

Proof of Theorem 6.2.2. Use the same construction as that in the proof of Theorem 4.1.1. For the spherical layers $L(r)$ with $R(r+1)\le n^{\alpha}/(\gamma+1)$, proof is word to word the same as in the mentioned theorem if we apply (6.17) and (4.2). Note that the total number of points θ in Θ_n has a power order of growth as $n\to\infty$. For r such that $rR > n^{\alpha}/(\gamma+1)$ we use Lemma 6.3.4 which guarantees that each of the probabilities $P_G\left(\hat{\theta}_{\ell}=\theta(m)\right)$ decreases as exponent in n for such $\theta(m)$ that $\|m-m_{\ell}^{(0)}\| > n^{\alpha}/(\gamma+1)$. In fact, as in (4.9), for n large enough we have

$$P_G\left(\hat{\theta}_{\ell}=\theta(m)\right) \le P_G\left(Z_{\ell}^*(\theta(m))-Z_{\ell}^*(\theta(m_{\ell}^{(0)})) > 0\right) \le$$

$$\le P_G\left(\Delta_{\ell}^*(\theta(m))-\Delta_{\ell}^*(\theta(m_{\ell}^{(0)})) > (1/4)\sum_{i\in I_{\ell}} P_G\left(X_i^*\in\mathscr{L}_i(\theta)\right)\right) \le$$

$$\le c_5\exp\left(\varkappa_0 z^2\sum_{i\in I_{\ell}} P_G\left(X_i^*\in\mathscr{L}_i(\theta)\right) - (z/4)\sum_{i\in I_{\ell}} P_G\left(X_i^*\in\mathscr{L}_i(\theta)\right)\right)$$

where $c_5 > 0$ is a constant and $\varkappa_0$ is defined by (6.14). If we take $z=(1/8\varkappa_0)$ and apply Lemma 6.3.4, then

$$P_G\left(\hat{\theta}_{\ell}=\theta(m)\right) \le c_5\exp\left(-\varkappa_1\sum_{i\in I_{\ell}} P_G\left(X_i^*\in\mathscr{L}_i(\theta)\right)\right) \le c_5\exp(-\varkappa_1 c_3 n^{\alpha})$$

where $\varkappa_1=(64\varkappa_0)^{-1}$. Hence,

$$P_G\left(\|\hat{m}_{\ell}-m_{\ell}^{(0)}\| > n^{\alpha}/(\gamma+1)\right) \le$$
$$\le c_5(\mathrm{card}\,\Theta_n)\exp(-\varkappa_1 c_3 n^{\alpha}) \le \exp(-\varkappa_2 n^{\alpha})$$

for $\varkappa_2=\varkappa_1 c_3$ and n large enough. The rest of proof is the same as in Theorem 4.1.1. □

6.4. LINEAR IMAGE ESTIMATORS

In this section we consider the noisy image model (3.1). Linear image estimators are statistics of the form

$$\mathcal{L}_n(x) = \sum_{i=1}^{n} Y_i \phi_{in}(x, X_1, \ldots, X_n)$$

where ϕ_{in} are arbitrary measurable functions.

It seems reasonable that for nonsmooth (discontinuous) images linear estimates are not the best ones. However, it is easily shown that convergence of MISE for linear estimates is not arbitrarily slow. For example, consider the step-function image

$$f(x) = I\{x=(x_1,x_2)\in[0,1]^2 : x_2 \le 1/2\}.$$

Assume that $\mathcal{L}_n$ is the simplest linear estimate of the form

$$\mathcal{L}_n(x) = m_n^{-1} \sum_{i\in\mathcal{I}_n} Y_i$$

where $\mathcal{I}_n = \mathcal{I}_n(x) = \{z\in[0,1]^2 : |z-x| \le h_n\}$, $m_n = \mathrm{card}\{i : X_i \in \mathcal{I}_n\}$, and $h_n \to 0$, $nh_n^2 \to \infty$ as $n\to\infty$. Under the regular design (A1) $m_n = O(nh_n^2)$ and the variance

$$\mathrm{Var}\Big(\mathcal{L}_n(x) - f(x)\Big) \le c(nh_n^2)^{-1}$$

for any fixed $x\in K$ where c does not depend on x. For $|x_2 - 1/2| \le h_n$ the values of the bias

$$f(x) - m_n^{-1} \sum_{i\in\mathcal{I}_n} f(X_i)$$

are less than 1, while for $|x_2-1/2| > h_n$ the bias equals to 0. Thus, the overall integrated risk is

$$\text{(6.19)} \qquad \mathrm{MISE}(\mathcal{L}_n, f) = O((nh_n^2)^{-1}) + O(h_n), \text{ as } n\to\infty.$$

The balance of two terms in the right-hand side of (6.19) gives the value $h_n = n^{-1/3}$. The rate of convergence is therefore equal to $n^{-1/6}$ which is slower than the minimax rate for binary image estimation. However, as follows from Theorem 6.4.1 given below this rate cannot be improved by linear estimates, at least on the classes of images $\Phi^0_{\infty,\gamma,N}$.

6.4.1. THEOREM. Let the design $(X_1,\dots,X_n)$ be arbitrary but independent of $(\xi_1,\dots,\xi_n)$, and let $E(\xi_i)=0$, $E(\xi_i^2)<\infty$. Then

$$\liminf_{n\to\infty} \inf_{\mathfrak{L}_n} \sup_{f\in\Phi^0_{\infty,\gamma,N}} \psi_n^{-2}\mathrm{MISE}(\mathfrak{L}_n,f) \ge C_0 > 0,\ \gamma>0,$$

where $\psi_n=n^{-\gamma/(2(2\gamma+N-1))}$ and $\inf_{\mathfrak{L}_n}$ is the infimum over all linear estimators.

<u>Proof</u>. Denote $\phi_i(x)=\phi_{in}(x,X_1,\dots,X_n)$,

$$\tau^2 = \sup_{f\in\Phi^0_{\infty,\gamma,N}} \mathrm{MISE}(\mathfrak{L}_n,f).$$

For any linear estimator $\mathfrak{L}_n$ we have

$$\mathrm{MISE}(\mathfrak{L}_n,f) = E\left(\int \sum_{i=1}^n \phi_i^2(z)\,dz + \int \sum_{i=1}^n (f(x_i)\phi_i(z)-f(z))^2 dz \right).$$

Hence

$$\tau^2 \ge \int \sum_{i=1}^n E(\phi_i^2(z))\,dz. \tag{6.20}$$

Let

$$\mathcal{A} = \left\{\mathcal{X}=(X_1,\dots,X_n): \int \sum_{i=1}^n \phi_i^2(z)\,dz \le 2\tau^2\right\}.$$

By (6.20) we have

$$P(\mathcal{A}) \ge 1/2. \tag{6.21}$$

Denote $f(x,\omega,r) = I\{x\in G^r(\bar\omega)\}$ where $G^r(\bar\omega)$ is defined in the proof of Theorem 5.1.1. Parameter m appearing in the definition of $G^r(\bar\omega)$ will be chosen later. Now we prove a modified version of (5.2). We use the notation from the proof of Theorem 5.1.1.

Let us show that for $\mathcal{X} = (X_1,\dots,X_n)\in\mathcal{A}$ fixed and for m large enough there exist a subset $D_{2m}=D_{2m}(\mathcal{X})$ of D_m and subsets $\Psi_{1r}=\Psi_{1r}(\mathcal{X})$ of $\{1,\dots,M\}$ such that

$$(6.22)\qquad \operatorname{card} U_{jr} \le 16n/((1-2h)m^{\gamma}M), \qquad \sum_{i\in U_{jr}} \int \phi_i^2(z)\,dz \le 32/(1-2h)(\tau^2/(m^{\gamma}M)),$$

and

$$(6.23)\qquad \operatorname{card} D_{2m} \ge (3/8)(1/2-h)m^{\gamma}, \qquad \operatorname{card} \Psi_{1r} \ge M/4.$$

Indeed, denote by D_{2m} the set of integers $r\in D_m$ which satisfy simultaneously (5.3) and the inequality

$$(6.24)\qquad \sum_{i:X_i\in\mathcal{S}_r} \int \phi_i^2(z)dz \le 8\tau^2/((1-2h)m^{\gamma}).$$

Since $\mathcal{X}\in\mathcal{A}$ there are no fewer than $(3/4+h/2)m^{\gamma}$ integers r satisfying (6.24). Hence there are no fewer than $(1/2+h)m^{\gamma}$ integers r that satisfy (5.3) and (6.24) simultaneously. Thus,

$$(6.25)\qquad \operatorname{card} D_{2m} \ge ((1/2+h)m^{\gamma}+\operatorname{card} D_m-m)/2 > (3/8)(1/2-h)m^{\gamma}$$

for $m^{\gamma} > 8/(1-2h)$. This gives the first inequality in (6.23). For each $r\in D_{2m}$, there are no fewer than $3M/4$ integers j such that the first inequality in (6.22) holds. In fact, otherwise we have

$$\operatorname{card}\{i:X_i\in\mathcal{S}_r\}\ge \sum_{j=1}^{M} \operatorname{card} U_{jr} > (M/4)(16/(1-2h))(n/m^{\gamma}M)=4n/((1-2h)m^{\gamma})$$

which contradicts (5.3). Similarly, there are no fewer than $3M/4$ integers j such that the second inequality in (6.22) is satisfied. Define Ψ_{1r} as the set of integers j that satisfy both inequalities in (6.22). Then it follows that for card Ψ_{1r} the second inequality in (6.23) holds.

As in (5.4) we get

$$(6.26)\qquad \tau^2 \ge (\text{card } D_m)^{-1}2^{-M}E\Big[I\{\mathcal{A}\}\sum_{r\in D_{2m}}\sum_{j\in\Psi_{1r}}$$

$$\int_{\mathcal{T}_{jr}}\Big\{\sum_{\omega\in\Omega}\Big(\sum_{i=1}^{n} f(X_i,\bar{\omega},r)\phi_i(z) - f(z,\bar{\omega},r)\Big)^2\Big\}dz\Big].$$

Here we used the fact that

$$E(\hat{f}_n(z)|\mathcal{X}) = \sum_{i=1}^{n} f(X_i,\bar{\omega},r)\phi_i(z).$$

Since $x^2+y^2 \ge (x-y)^2/2$ we obtain that for $\omega_1,\ldots,\omega_{j-1},\omega_{j+1},\ldots,\omega_M$ fixed

$$\sum_{\omega_j=0,1}\Big(\sum_{i=1}^{n} f(X_i,\bar{\omega},r)\phi_i(z) - f(z,\bar{\omega},r)\Big)^2 \ge$$

$$\ge (1/2)\Big(\sum_{i\in U_{jr}}\phi_i(z) - I\{z\in\mathcal{T}_{jr}\}\Big)^2.$$

Note that the last expression does not depend on $\omega_1,\ldots,\omega_{j-1},\omega_{j+1},\ldots,\omega_M$. By substitution of this inequality into (6.26) we get

$$(6.27)\qquad \tau^2 \ge (4\text{ card } D_m)^{-1}E\Big[I\{\mathcal{A}\}\sum_{r\in D_{2m}}\sum_{j\in\Psi_{1r}}\int_{\mathcal{T}_{jr}}\Big(\sum_{i\in U_{jr}}\phi_i(z)-1\Big)^2dz\Big].$$

By (6.22)

$$\Big[\int_{\mathcal{T}_{jr}}\Big(\sum_{i\in U_{jr}}\phi_i(z) - 1\Big)^2dz\Big]^{1/2} \ge$$

$$\ge \Big(\text{mes } \mathcal{T}_{jr}\Big)^{1/2} - \Big[\int_{\mathcal{T}_{jr}}\Big(\sum_{i\in U_{jr}}\phi_i(z)\Big)^2dz\Big]^{1/2} \ge$$

$$\geq c_6\left((m^\gamma M)^{-1/2} - \left[\text{card } U_{jr} \int \sum_{i\in U_{jr}} \phi_i^2(z)dz\right]^{1/2}\right) \geq$$

$$\geq c_7\left((m^\gamma M)^{-1/2} - \tau n^{1/2}/(m^\gamma M)\right)$$

where $c_6, c_7>0$ are constants. This together with (6.21), (6.23), (6.27) entails

$$\tau^2 \geq c_8 M\left((m^\gamma M)^{-1/2} - \tau n^{-1/2}/(m^\gamma M)\right)^2$$

with some $c_8>0$. Remind that $M=m^{N-1}$. Choosing $m=[n^{1/(2\gamma+N-1)}]$ we obtain the result of the theorem. □

For $\gamma=1$, $N=2$ the order of the lower bound of Theorem 6.4.1 coincides with that of the upper bound (6.19). It is easy to prove that such upper bound holds not only for step-function images but also for any images of the class $\Phi^0_{\infty,1,N}$ (Lipschitz edge binary boundary fragments). Thus the rate $\psi_n=n^{-1/6}$ is optimal <u>for linear estimators</u> on the class $\Phi^0_{\infty,1,N}$.

Another simple consequence of Theorem 6.4.1 is the lower bound for linear estimators on the classes $\Phi^0_{\beta,\gamma,N}$, $0<\beta<\infty$. It follows that such a bound holds with

$$\psi_n = \max\left\{(1/n)^{\gamma/(2(2\gamma+N-1))},\ (1/n)^{\beta/(2\beta+N)}\right\}.$$

Comparing this with the upper bound of Theorem 5.1.8 we get the following.

6.4.2. COROLLARY. Linear estimates are not S-optimal for the classes of boundary fragments $\Phi^0_{\beta,\gamma,N}$ in L_2-norm if $\gamma\geq 1$, $\gamma N/(2(\gamma+N-1)) \leq \beta \leq \infty$.

CHAPTER 7. ESTIMATION OF SUPPORT OF A DENSITY

7.1. PROBLEM STATEMENT

Let $X_1,\ldots,X_n$ be i.i.d. random variables uniformly distributed in a closed set $G \subset K=[0,1]^N$, mes $G \neq 0$. In this chapter we study the problem of estimation of G, given the observations $X_1,\ldots,X_n$. This problem is closely related to edge estimation, and, as we show later, it can be solved by similar means. The minimax convergence rates for estimators of supports with smooth boundaries turn out to be the same as those for edge estimators (cf. Chapters 4,5).

The support estimation problem was first considered in the papers of Geffroy (1964) and Rényi and Sulanke (1963, 1964). Geffroy (1964) studied the uniform convergence and the asymptotic distribution of maximal deviation of a piecewise-constant support estimator for two-dimensional boundary fragments. Rényi and Sulanke (1963, 1964) investigated the case of convex support G in two dimensions, and they proposed a natural estimator which is a convex hull of sample points. Their results were generalized to multidimensional case and extended in several ways (see the survey of Schneider (1988) and references therein, Bosq (1971), Chevalier (1976), Moore (1984)).

The support estimation is useful for several applications. We mention here some of them. First, as proposed by Devroye and Wise (1980), one can apply support estimation for a kind of classification scheme to detect the abnormal behaviour of a system, plant or machine. If $\hat{G}_n$, the estimator of support G, is given then one can check whether the new observation $X\in K$ is abnormal, i.e. whether X is not driven from the same population as $X_1,\ldots,X_n$. The system is decided to behave abnormally if $X\notin\hat{G}_n$.

The second application of support estimation is cluster analysis. One can define clusters as level sets of densities, i.e. the sets where the density f(x) of X_i's is larger than some constant $\lambda>0$ (see Hartigan (1975)). In our case of uniform density f(x) all the level sets $V_\lambda=\{f(x)\geq\lambda\}$ coincide with support G. Note that the assumption of uniform density is a simplification introduced here to make the relation to edge estimation more transparent. If the density is not uniform then also its smoothness properties are important, and they influence the error of level set estimation. The interesting problem of optimal rates in level set estimation for general smooth densities is not yet solved. Some upper bounds for the risks are given by Hartigan (1987) and Polonik (1991).

The related problem is excess mass estimation (Müller and Sawitzky (1987,1991), Hartigan (1987), Polonik (1991)). The excess mass is defined as $\int \max\{f(x)-\lambda, 0\}dx$. If $f(x)$ is the uniform density on G and $\lambda<(\text{mes } G)^{-1}$, then the excess mass equals mes G times a constant, and thus the problem of excess mass estimation reduces to estimation of the area of support G.

In econometrics the estimation of the boundary of density support allows to evaluate the performance of an enterprise in terms of technical efficiency. The latter is defined as the distance from the observed productivity to the boundary. Here the support G is supposed to be of multivariate boundary fragment type. The first N-1 components of X_i describe the input parameters of the ith enterprise, and the Nth component is its productivity. Nonparametric methods of estimation of boundary were proposed by Farrell (1957) and Deprins, Simar and Tulkens (1984). The "data envelopment analysis" of Farrell (1957) suggests the convex hull of data points as support estimator, while Deprins, Simar and Tulkens (1984) introduce the "free-disposal hull" which is the monotone boundary estimator.

It is also worthwhile to mention the problem of finding the edge of Poisson forest which shows much resemblance to estimation of support of a density. This problem posed by D.G.Kendall is the following: given a realization of a homogeneous planar Poisson point process of unknown intensity within a compact set G, estimate G. For the case of convex G this problem was studied by Ripley and Rasson (1977) and for some other types of domains by Jacob (1984), Jacob and Abbar (1989) and Abbar (1990).

This chapter is organized as follows. In Section 7.2 we study a simple and rough support estimator. In Section 7.3 the proof of minimax lower bounds is given for support estimation on classes of boundary fragments. Sections 7.4, 7.5 are devoted to the optimal support estimators.

7.2. A SIMPLE AND ROUGH SUPPORT ESTIMATOR

The estimator that we analyse here is the analogue to simple and rough edge estimator of Section 5.2. It was proposed by Chevalier (1976) and Devroye and Wise (1980). Chevalier (1976) studied its convergence and rate of convergence for the case of ε-convex sets. Devroye and Wise (1980) proved its consistency in d_1-metric under very general conditions. Here we show that this estimator has a certain convergence rate which appears to be the same as that of the edge estimator G_n^{**} of Section 5.2.

7.2.1. DEFINITION. The support estimator G_n^{**} is the union of balls of radius $h_n>0$ around the sample points:

$$G_n^{**} = \bigcup_{i=1}^{n} \{x\in K: |X_i - x| \le h_n\}.$$

Here h_n's are bandwidths satisfying the usual conditions $h_n\to 0$, $nh_n^N\to\infty$ as $n\to\infty$ (cf. Chapter 1).

Consider the class $\mathcal{G}_N$ of domains with piecewise Lipschitz boundaries defined in 5.2.2. The following theorem gives the rate of convergence of G_n^{**} on the class $\mathcal{G}_N$ in d_1-metric.

7.2.2. THEOREM. Let $h_n \sim (\tilde{C} \log n/n)^{1/N}$, $n\to\infty$. If $\tilde{C}>0$ is large enough, then the estimator G_n^{**} satisfies

$$\sup_{G\in\mathcal{G}_N} E_G(d_1(G,G_n^{**})) = O((\log n/n)^{1/N}),\ n\to\infty.$$

Here and later in this chapter E_G denotes the expectation over the distribution P_G of $X_1,\dots,X_n$ where X_i's are uniformly distributed in G.

Proof. Denote $B_i=\{x\in K: |X_i-x|\le h_n\}$, $Q = \{x\in K: \rho(x,\partial G)\le h_n\}$ where $\rho(x,\partial G)$ is Euclidean distance between the point x and the boundary ∂G of G. Clearly,

$$G\Delta G_n^{**} \subseteq (G\backslash G_n^{**}) \cup Q.$$

Hence

$$(7.1)\qquad E_G(d_1(G,G_n^{**})) \le E_G(\text{mes}(G\backslash G_n^{**})) + \text{mes}\, Q =$$

$$= E_G\int I\{x\in G\backslash G_n^{**}\}dx + \text{mes}\, Q = \int P_G(x\in G\backslash G_n^{**})dx + \text{mes}\, Q.$$

Assumption 5.2.2(i) implies

$$(7.2)\qquad \text{mes}\, Q = \mathcal{O}(h_n).$$

Due to the independence of X_i's

$$P_G(x\in G\backslash G_n^{**}) = P_G(\bigcap_{i=1}^{n}\{x\in G\backslash B_i\}) = \prod_{i=1}^{n}\Big[1 - P_G(x\in G\cap B_i)\Big].$$

Since, by Assumption 5.2.2(ii), G satisfies (a_1,a_2)-cone condition and the center X_i of B_i belongs to G then $G\cap B_i$ contains a spherical sector of fixed angle and of radius h_n (if n is large enough, so that $a_2>h_n$). Hence $\text{mes}(G\cap B_i) \ge ch_n^N \ge (c/2)\tilde{C}(\log n/n)$ for n large enough where $c>0$ depends only on a_1, N. This gives

$$P_G(x\in G\backslash G_n^{**}) \le (1-(c/2)\tilde{C}(\log n/n))^n \le \exp(-(c/2)\tilde{C}(\log n)).$$

Choose $\tilde{C}$ to satisfy the condition $(c/2)\tilde{C} \ge 1/N$. Then

$$P_G(x\in G\backslash G_n^{**}) = \mathcal{O}(n^{-1/N}),$$

and the theorem follows from (7.1), (7.2). □

7.3. MINIMAX LOWER BOUNDS FOR SUPPORT ESTIMATION

In this section we derive the lower bounds for minimax risks of support estimation. It suffices to obtain lower bounds for narrowest classes of domains, namely, for boundary fragments. We consider here the classes $\mathcal{G}_{0,N}$ of N-dimensional boundary fragments defined in Section 5.1. The proof of lower bounds is based on Assouad's lemma (Theorem 2.6.4), and is similar to the proof of Theorem 5.1.1.

7.3.1. THEOREM. For the support estimation problem we have

$$\liminf_{n\to\infty} \inf_{\hat{G}_n} \sup_{G\in\mathcal{G}_{0,N}} E_G(\psi_n^{-q} d_1^q(G,\hat{G}_n)) \ge C_0 > 0, \ \forall q\ge 1,$$

where $\psi_n = n^{-\gamma/(\gamma+N-1)}$. Here C_0 depends only on γ, N, L, q.

Proof. It suffices to consider the case $q=1$ (see the proof of Theorem 5.1.1). Consider the family of edge functions defined in the proof of Theorem 5.1.1, and put there $r=(1/2)m^{\gamma}$. Thus, we get the following edge functions

$$g(t,\bar{\omega}) = 1/2 + (\lambda/m^{\gamma}) \sum_{j=1}^{M} \omega_j \phi(m(t-b_j)),$$

$$m=[n^{1/(\gamma+N-1)}], \ M=m^{N-1}, \ \bar{\omega} = (\omega_1,\ldots,\omega_M), \ \omega_j\in\{0,1\},$$

$$t=(x_1,\ldots,x_{N-1}).$$

Define the domains

$$G(\bar{\omega}) = \left\{ x\in K: 0 \le x_N \le g((x_1,\ldots,x_{N-1}),\bar{\omega}) \right\}.$$

We have $g(\cdot,\bar{\omega})\in\Sigma_{N-1}(\gamma,L,h)$ for λ small enough, and thus $G(\bar{\omega})\in\mathcal{G}_{0,N}$. Clearly,

$$\min_{\bar{\omega}} \text{mes } G(\bar{\omega}) \ge 1/2. \tag{7.3}$$

Denote by $P_{i,G}$ the distribution of X_i when the underlying domain is G. As in Theorems 2.6.4, 5.1.1 introduce the vectors

$$\bar{\omega}_{j0} = (\omega_1,\ldots,\omega_{j-1},0,\omega_{j+1},\ldots,\omega_M),$$

$$\bar{\omega}_{j1} = (\omega_1,\ldots,\omega_{j-1},1,\omega_{j+1},\ldots,\omega_M).$$

Then

$$(7.4)\qquad H^2\left(P_{i,G(\bar{\omega}_{j0})}, P_{i,G(\bar{\omega}_{j1})}\right) =$$

$$= \int \left(\sqrt{\frac{I\{x \in G(\bar{\omega}_{j0})\}}{\text{mes } G(\bar{\omega}_{j0})}} - \sqrt{\frac{I\{x \in G(\bar{\omega}_{j1})\}}{\text{mes } G(\bar{\omega}_{j1})}} \right)^2 dx =$$

$$= 2 - 2\int \left(\frac{I\{x \in G(\bar{\omega}_{j0})\} I\{x \in G(\bar{\omega}_{j1})\}}{\text{mes } G(\bar{\omega}_{j0}) \text{ mes } G(\bar{\omega}_{j1})} \right)^{1/2} dx =$$

$$= 2\left(1 - \int \frac{I\{x \in G(\bar{\omega}_{j0})\}}{(\text{mes } G(\bar{\omega}_{j0}) \text{ mes } G(\bar{\omega}_{j1}))^{1/2}} dx \right) \le$$

$$\le 2\sqrt{2}\left((\text{mes } G(\bar{\omega}_{j1}))^{1/2} - (\text{mes } G(\bar{\omega}_{j0}))^{1/2} \right)$$

where we used (7.3). The definition of $G(\bar{\omega}_{j0})$, $G(\bar{\omega}_{j1})$ entails

$$(7.5)\qquad \begin{aligned} &d_1(G(\bar{\omega}_{j0}), G(\bar{\omega}_{j1})) = \text{mes}(G(\bar{\omega}_{j0}) \Delta G(\bar{\omega}_{j1})) = C_{mes}/n, \\ &\text{mes}(G(\bar{\omega}_{j1})) = \text{mes}(G(\bar{\omega}_{j0})) + \text{mes}(G(\bar{\omega}_{j0}) \Delta G(\bar{\omega}_{j1})) = \\ &= \text{mes}(G(\bar{\omega}_{j0})) + C_{mes}/n \end{aligned}$$

where C_{mes} is a positive constant. It follows from (7.3), (7.5) that for n large

$$(\text{mes } G(\bar{\omega}_{j1}))^{1/2} - (\text{mes } G(\bar{\omega}_{j0}))^{1/2} \le C_{mes}/n$$

(note that $(a+b)^{1/2} - a^{1/2} \le b$ if $a \ge 1/2$, $0 \le b < 1$). Hence (7.4) reduces to

$$(7.6)\qquad H^2\left(P_{i,G(\bar{\omega}_{j0})}, P_{i,G(\bar{\omega}_{j1})}\right) \le 2\sqrt{2}C_{mes}/n.$$

As in (2.45) - (2.47) and (5.4) - (5.6), one gets

$$\sup_{G\in\mathcal{G}_{0,N}} E_G(d_1(G,\hat{G}_n)) \geq$$

$$\geq 2^{-M}\sum_{j=1}^{M}\sum{}'\Big[E_{G(\bar{\omega}_{j0})}\Big(d_1(G(\bar{\omega}_{j0})\cap A_j,\hat{G}_n\cap A_j)\Big) +$$

$$+ E_{G(\bar{\omega}_{j1})}\Big(d_1(G(\bar{\omega}_{j1})\cap A_j,\hat{G}_n\cap A_j)\Big)\Big] \geq$$

$$\geq 2^{-1}\sum_{j=1}^{M} d_1(G(\bar{\omega}_{j0}),G(\bar{\omega}_{j1}))\int\min\Big(\prod_{i=1}^{n} dP_{i,G(\bar{\omega}_{j0})},\prod_{i=1}^{n} dP_{i,G(\bar{\omega}_{j1})}\Big) \geq$$

$$\geq (C_{mes}n^{-1}/4)\sum_{j=1}^{M}\prod_{i=1}^{n}\Big(1 - H^2\Big(P_{i,G(\bar{\omega}_{j0})},P_{i,G(\bar{\omega}_{j1})}\Big)/2\Big)^2 \geq$$

$$\geq (C_{mes}Mn^{-1}/4)(1 - \sqrt{2}C_{mes}/n)^{2n}$$

where we used (7.5), (7.6). This, in view of the definition of M, entails the theorem. □

7.4. OPTIMAL SUPPORT ESTIMATION FOR BOUNDARY FRAGMENTS

The construction of optimal edge estimator in Section 4.1 can be applied with necessary modifications to the support estimation problem. Remind that in Section 4.1 we used the piecewise-polynomial estimator, with the coefficients of polynomials obtained by discretized maximum likelihood procedure. Here we apply the same method for support estimation. The only difference appears in the structure of the likelihood function.

First, we prove optimality of the estimators in the two-dimensional case (N=2) keeping the notation of Sections 4.1-4.3. Next, we state the result for general case of $N\geq 2$. The proof for the general case is not given since it is quite similar but requires more bulky notation.

Assume that the support G belongs to the class of two-dimensional boundary fragments $\mathcal{G}_0=\mathcal{G}_0(\gamma, L, h)$. In each slice $A_\ell=[u_{\ell-1}, u_\ell)\times R^1$, $\ell=1,\ldots,M$, define the discretized maximum likelihood estimator $\hat{\theta}_\ell=(\hat{\theta}_{0\ell},\ldots,\hat{\theta}_{k\ell})$ as

$$(7.7) \qquad \hat{\theta}_\ell = \underset{\theta\in\Theta_n}{\operatorname{argmax}} \prod_{i:X_i\in A_\ell} \left[\frac{I\{X_i\in B_\ell(\theta)\}}{\text{mes } B_\ell(\theta)} \right] =$$

$$= \underset{\substack{\theta\in\Theta_n: \\ B_\ell(\theta)\supseteq\{X_i: X_i\in A_\ell\}}}{\operatorname{argmin}} \text{mes } B_\ell(\theta)$$

where $B_\ell(\theta)$ is the parametric domain with polynomial edge and Θ_n is the discrete set of parameters (see Section 4.1 for definitions). Assume that the bin width is

$$\delta_n = \begin{cases} (1/n)^{1/(\gamma+1)} & \text{if } d=d_1, \\ (\log n/n)^{1/(\gamma+1)} & \text{if } d=d_\infty. \end{cases}$$

The piecewise-polynomial support estimator G_n^* is defined as the closure of the set

$$\left\{(x_1,x_2):\ 0 \leq x_2 \leq \hat{g}(x_1),\ 0 \leq x_1 \leq 1\right\}$$

where

$$\hat{g}(x_1) = \hat{g}_\ell(x_1) = \hat{\theta}_{0\ell}+\hat{\theta}_{1\ell}(x_1-u_{\ell-1})+\ldots+\hat{\theta}_{k\ell}(x_1-u_{\ell-1})^k,$$

$$u_{\ell-1} \leq x_1 < u_\ell,\ \ell=1,\ldots,M.$$

7.4.1. THEOREM. The piecewise-polynomial support estimator G_n^* satisfies

$$\limsup_{n\to\infty} \sup_{G\in\mathcal{G}_0} E_G(\psi_n^{-q} d^q(G,G_n^*)) < \infty, \; q>0,$$

where

(i) $\psi_n=(1/n)^{\gamma/(\gamma+1)}$ if d is the measure of symmetric difference d_1;

(ii) $\psi_n=(\log n/n)^{\gamma/(\gamma+1)}$ if d is the Hausdorff distance d_∞.

<u>Proof.</u> Let $g_\ell^{(0)}(x_1)$ be the Taylor polynomial with discretized coefficients as defined in Section 4.2. Since $g\in\Sigma(\gamma,L)$, we have

$$|g(x_1) - g_\ell^{(0)}(x_1)| \le \delta_n^\gamma L/k! + (k+1)\delta_n^\gamma$$

where the last summand appears because of discretization. Thus, the polynomial

$$g_\ell^{(1)}(x_1) = g_\ell^{(0)}(x_1) + \delta_n^\gamma(\lfloor L/k! + (k+1)\rfloor + 1)$$

satisfies

$$g_\ell^{(1)}(x_1) \ge g(x_1).$$

Clearly, $\theta_\ell^{(1)}\in\Theta_n$ where $\theta_\ell^{(1)}$ is the vector of coefficients of $g_\ell^{(1)}(x_1)$, and in view of the last inequality the set $B_\ell(\theta_\ell^{(1)})$ satisfies

$$B_\ell(\theta_\ell^{(1)}) \supseteq D_\ell \supset \{X_i:\; X_i\in A_\ell\}. \tag{7.8}$$

Let $m_\ell^{(1)}$ be the vector of integers corresponding to $\theta_\ell^{(1)}$, i.e.

$$m_\ell^{(1)} = (m^{(0,0)}+\lfloor L/k! + (k+1)\rfloor + 1,\; m^{(0,1)},\dots,\; m^{(0,k)})$$

where $m^{(0,j)}$ are defined in Section 4.2.

For any $R>0$ we have

$$(7.9) \qquad P_G\left(\sup_{u_{\ell-1}\le x_1<u_\ell} |\hat{g}(x_1)-g(x_1)| \ge R\delta_n^\gamma \right) \le$$

$$\le P_G\left(\sup_{u_{\ell-1}\le x_1<u_\ell} |\hat{g}(x_1)-g_\ell^{(1)}(x_1)| + \sup_{u_{\ell-1}\le x_1<u_\ell} |g(x_1)-g_\ell^{(1)}(x_1)| \ge R\delta_n^\gamma \right) \le$$

$$\le P_G\left(\sup_{u_{\ell-1}\le x_1<u_\ell} |\hat{g}(x_1)-g_\ell^{(1)}(x_1)| \ge R\delta_n^\gamma - 2\delta_n^\gamma(L/k! + k+2) \right).$$

Clearly,

$$\sup_{u_{\ell-1}\le x_1<u_\ell} |\hat{g}(x_1)-g_\ell^{(1)}(x_1)| \le (k+1)\delta_n^{\gamma}\|\hat{m}_\ell-m_\ell^{(1)}\|.$$

By Lemma 4.2.1

$$\|\hat{m}_\ell-m_\ell^{(1)}\| \le \kappa_1^{-1}\delta_n^{-\gamma-1}d_1(B_\ell(\hat{\theta}_\ell), B_\ell(\theta_\ell^{(1)})).$$

Hence

$$\sup_{u_{\ell-1}\le x_1<u_\ell} |\hat{g}(x_1)-g_\ell^{(1)}(x_1)| \le (k+1)\kappa_1^{-1}\delta_n^{-1}d_1(B_\ell(\hat{\theta}_\ell), B_\ell(\theta_\ell^{(1)})).$$

By substitution of this inequality into (7.9) we get

$$(7.10) \qquad P_G\left(\sup_{u_{\ell-1}\le x_1<u_\ell} |\hat{g}(x_1)-g(x_1)| \ge R\delta_n^\gamma \right) \le$$

$$\le P_G\left(d_1(B_\ell(\hat{\theta}_\ell), B_\ell(\theta_\ell^{(1)})) \ge \delta_n^{\gamma+1}\kappa_1(k+1)^{-1}(R-2(L/k!+k+2)) \right).$$

By definition of $\hat{\theta}_\ell$

$$(7.11) \qquad d_1(B_\ell(\hat{\theta}_\ell), B_\ell(\theta_\ell^{(1)})) \le \text{mes}\,(B_\ell(\theta_\ell^{(1)})\setminus B_\ell(\hat{\theta}_\ell)).$$

It follows from (7.10), (7.11) that there exists $c>0$ such that for R large enough

$$(7.12)\qquad P_G\left(\sup_{u_{\ell-1}\le x_1<u_\ell} |\hat{g}(x_1)-g(x_1)| \ge R\delta_n^{\gamma}\right) \le$$

$$\le P_G\left(\text{mes }(B_\ell(\theta_\ell^{(1)})\setminus B_\ell(\hat{\theta}_\ell)) \ge cR\delta_n^{\gamma+1}\right) =$$

$$= P_G\left(\text{mes }[B_\ell(\theta(m_\ell^{(1)}))\setminus B_\ell(\theta(\hat{m}_\ell))] \ge cR\delta_n^{\gamma+1};\right.$$

$$\left. X_i\notin D_\ell\setminus B_\ell(\theta(\hat{m}_\ell)),\ i=1,\dots,n\right)$$

where $\theta(m_\ell^{(1)})=\theta_\ell^{(1)}$, $\theta(\hat{m}_\ell)=\hat{\theta}_\ell$ according to the notation of Section 4.2, and the last equality follows from (7.8) and the fact that $X_i\notin A_\ell\setminus B_\ell(\theta(\hat{m}_\ell))$, $i=1,\dots,n$.

Now we act as in the proof of Theorem 4.1.1. The last probability in (7.12) does not exceed

$$(7.13)\qquad \sum_{r\ge1} \text{card } L'(r) \max_{m\in L'(r)} P_G\left(X_i\notin D_\ell\setminus B_\ell(\theta(m)),\ i=1,\dots,n\right)$$

where

$$L'(r)=L(r)\cap\left\{ m\colon \text{mes }[B_\ell(\theta(m_\ell^{(1)}))\setminus B_\ell(\theta(m))] \ge cR\delta_n^{\gamma+1} \right\}.$$

For $m\in L'(r)$ we have

$$\text{mes}(D_\ell\setminus B_\ell(\theta(m))) \ge \text{mes}[B_\ell(\theta(m_\ell^{(1)}))\setminus B_\ell(\theta(m))] -$$

$$- \text{mes}[B_\ell(\theta(m_\ell^{(1)}))\setminus D_\ell] \ge rcR\delta_n^{\gamma+1} - 2\delta_n^{\gamma+1}(L/k!+k+2) \ge rc'R\delta_n^{\gamma+1}$$

for R large enough. Here $c'>0$ is a constant. Hence

$$(7.14) \qquad P_G\left(X_i \notin D_\ell \backslash B_\ell(\theta(m)),\ i=1,\dots,n \right) =$$

$$= \left(1 - \operatorname{mes}(D_\ell \backslash B_\ell(\theta(m))) \right)^n \le$$

$$\le \left(1 - rc'R\delta_n^{\gamma+1} \right)^n, \quad m \in L'(r).$$

Assume that $d=d_1$. Then $\delta_n = n^{-1/(\gamma+1)}$, and (7.14) does not exceed $\exp(-c'Rr)$. Since card $L(r) = O(r^k R^{k+1})$ (cf. Section 4.3) we find that the sum in (7.13) does not exceed $\exp(-c''R)$ where $c''>0$ is a constant independent of ℓ and g. Thus, if R is large, then for each $\ell=1,\dots,M$,

$$P_G\left(\sup_{u_{\ell-1} \le x_1 < u_\ell} |\hat{g}(x_1) - g(x_1)| \ge R\delta_n^\gamma \right) \le \exp(-c''R)$$

This entails

$$E_G\left(\sup_{u_{\ell-1} \le x_1 < u_\ell} |\hat{g}(x_1) - g(x_1)|^q \delta_n^{-\gamma q} \right) = O(1), \quad n\to\infty, \quad q>0,$$

uniformly in $\ell=1,\dots,M$. Therefore,

$$E_G\left(\int_0^1 |\hat{g}(x_1) - g(x_1)|^q dx_1 \ \delta_n^{-\gamma q} \right) = O(1), \quad n\to\infty, \quad q>0,$$

where $O(1)$ does not depend on g. This proves the theorem in the case of $d=d_1$.

If $d=d_\infty$, then (7.14) is smaller than $n^{-c'Rr}$. This leads to

$$(7.15) \qquad P_G\left(\sup_{u_{\ell-1} \le x_1 < u_\ell} |\hat{g}(x_1) - g(x_1)| \ge R\delta_n^\gamma \right) \le n^{-R\tilde{c}}$$

for some $\tilde{c}>0$. The rest of the proof follows as in Theorem 4.1.1 (ii). □

Consider now the extension of Theorem 7.4.1 to the case of arbitrary $N\ge 2$. We use the notation introduced in Section 5.1. In each cube Q_ℓ, $\ell=1,\dots,m_n^{N-1}$, define the estimator $\hat{g}(x_1,\dots,x_{N-1})$ of the edge function $g(x_1,\dots,x_{N-1})$ as a polynomial $p(x_1,\dots,x_{N-1})$ of degree k (k is the maximal integer such that $k<\gamma$) solving the following minimization problem

$$\min_{p\in\mathcal{P}_{N,k,\ell}} \int p(x_1,\dots,x_{N-1})\, dx_1\dots dx_{N-1}. \tag{7.16}$$

Here

$$m_n = \begin{cases} [n^{1/(\gamma+N-1)}] & \text{if } d=d_1, \\ [(n/\log n)^{1/(\gamma+N-1)}] & \text{if } d=d_\infty, \end{cases}$$

and $\mathcal{P}_{N,k,\ell}$ is the finite set of polynomials:

$$\mathcal{P}_{N,k,\ell} = \mathcal{P}_{N,k} \cap \Big\{ p:\ X_{i(N)} \le p(X_{i(1)},\dots,X_{i(N-1)}),\quad X_i = (X_{i(1)},\dots,X_{i(N)})\in A_\ell \Big\}.$$

(for the definition of $\mathcal{P}_{N,k}$ see Section 5.1). The estimator G_n^* is defined as the closure of the set

$$\Big\{(x_1,\dots,x_N):\ 0 \le x_N \le \hat{g}(x_1,\dots,x_{N-1}),\ 0\le x_1,\dots,x_N\le 1\Big\}. \tag{7.17}$$

7.4.2. THEOREM. The support estimator G_n^* which is the closure of the set (7.17) satisfies

$$\limsup_{n\to\infty}\ \sup_{G\in\mathcal{G}_{0,N}} E_G(\psi_n^{-q} d^q(G,G_n^*)) < \infty,\quad q>0,$$

where

(i) $\psi_n=(1/n)^{\gamma/(\gamma+N-1)}$ if d is the measure of symmetric difference d_1;

(ii) $\psi_n=(\log n/n)^{\gamma/(\gamma+N-1)}$ if d is the Hausdorff distance d_∞.

Note that, unlike the related edge estimators of Section 4.1, the support estimators discussed here are easily calculable. It is straightforward to compute $\hat{\theta}_\ell$ from (7.7) or (7.16) even without the discretization of parametric space. In fact, the polynomial $\hat{g}$ is linear in $\hat{\theta}_\ell$ for $u_{\ell-1} \le x_1 \le u_\ell$, and thus the minimization problem in (7.7) reduces to standard linear programming with inequality restrictions. The same is true for the problem (7.16).

7.5. OPTIMAL SUPPORT ESTIMATION FOR CONVEX DOMAINS AND FOR DUDLEY'S CLASSES

Let the support G of the density of X_i's belong to the Dudley class $\tilde{\mathcal{G}}$ defined in Section 5.4. Consider the maximum likelihood estimator on ε-net that has the form

$$G_n^* = \underset{G \in \mathcal{N}_\varepsilon}{\operatorname{argmax}} \prod_{i=1}^{n} (\text{mes } G)^{-1} I\{X_i \in G\}$$

where $\mathcal{N}_\varepsilon$ is a minimal ε-net on $\tilde{\mathcal{G}}$ in d_1-metric. One can write G_n^* in the equivalent form:

$$G_n^* = \underset{\substack{G \in \mathcal{N}_\varepsilon: \\ G \supseteq \{X_1, \ldots, X_n\}}}{\operatorname{argmin}} \text{mes } G. \tag{7.18}$$

The estimator (7.18) is a discretized version of maximum likelihood estimator

$$G_n^{MLE} = \underset{\substack{G \in \tilde{\mathcal{G}}: \\ G \supseteq \{X_1, \ldots, X_n\}}}{\operatorname{argmin}} \text{mes } G.$$

The study of convergence rates of G_n^{MLE} is more involved than that of the estimator (7.18), and it falls beyond the scope of this book. We refer to the general result of Birgé and Massart (1991) on convergence rates of the minimum contrast estimators. An important example of the maximum likelihood support estimator has been considered by Rényi and Sulanke (1963, 1964). They studied the problem of support estimation in the case of convex domains G. The maximum likelihood estimator is then the convex hull of $X_1, \ldots, X_n$:

$$(7.19)\qquad G_n^{MLE} = \underset{\substack{G\in\mathcal{G}_{conv}:\\ G\supseteq\{X_1,\dots,X_n\}}}{\operatorname{argmin}}\ \text{mes}\, G = \text{conv}\{X_1,\dots,X_n\}.$$

Rényi and Sulanke (1964) assumed that the edge ∂G is twice continuously differentiable and proved that for dimension N=2 the estimator (7.19) satisfies

$$(7.20)\qquad \text{mes}\, G - E_G(\text{mes}\, G_n^{MLE}) = O(n^{-2/3}),\ n\to\infty.$$

Since G is convex we have $G_n^{MLE} \subseteq G$. Hence (7.20) implies

$$E_G(d_1(G_n^{MLE}, G)) = O(n^{-2/3}),\ n\to\infty.$$

On the other hand, the minimax lower bound for support estimation on the class of convex domains $\mathcal{G}_{conv}$ holds with the rate $\psi_n=n^{-2/(N+1)}$ (this can be shown by a modification of the proof of Theorem 7.3.1, see Mammen and Tsybakov (1992) for details). This entails that the convex hull of sample points is optimal estimator of convex domains for the case N=2. The generalization of this statement to the case of N>2 requires the upper bound of the order $O(n^{-2/(N+1)})$ on $E_G(d_1(G_n^{MLE},G)) = \text{mes}\, G - E_G(\text{mes}\, G_n^{MLE})$. A remark in Schneider (1988), p.223, shows that such a bound can be obtained as a combination of known results. Thus, the convex hull of sample points is A-optimal estimator of density support for the class $\mathcal{G}_{conv}$ in d_1-metric. Mammen and Tsybakov (1992) show that the optimal rate $n^{-2/(N+1)}$ is achieved by the maximum likelihood estimator on ε-net not only with respect to the absolute value loss, but for any power loss function (see Theorem 7.5.3 below).

7.5.1. REMARK. If the estimation of <u>domain's area</u> is considered (which was the original problem of Rényi and Sulanke (1964)), then the estimator $\text{mes}(G_n^{MLE})$ is not optimal. For convex domains it is possible to achieve the faster rate $n^{-5/6}$ (see Section 8.5.6). Note also that the convex hull estimator (7.19) is obviously "inwards" biased since $\text{conv}\{X_1,\dots,X_n\}\subset G$ almost surely.

7.5.2. THEOREM. Let $\mathcal{G}$ be a subset of $\tilde{\mathcal{G}}$ such that

$$\inf_{G\in\mathcal{G}} \text{mes}\, G > 0. \tag{7.21}$$

Then there exists an ε-net $\mathcal{N}_\varepsilon$ on $\mathcal{G}$ in d_1-metric, with $\varepsilon=\varepsilon^*=cn^{-\gamma/(\gamma+N-1)}$, $c>0$, such that the estimator G_n^* defined by (7.18) satisfies

$$\sup_{G\in\mathcal{G}} E_G(\psi_n^{-q} d_1^q(G,G_n^*)) = O(1),\ n\to\infty,\ \forall q>0, \tag{7.22}$$

where $\psi_n=n^{-\gamma/(\gamma+N-1)}$.

Proof of this theorem is based on a modification of Lemma 5.5.3, and it is given in Mammen and Tsybakov (1992), as well as the following analogue of Theorem 7.5.2 for the class of convex domains:

7.5.3. THEOREM. Let $\mathcal{G}$ be a subset of $\mathcal{G}_{\text{conv}}$ satisfying (7.21). Then there exists an ε-net $\mathcal{N}_\varepsilon$ on $\mathcal{G}$ in d_1-metric, with $\varepsilon=\varepsilon^*=cn^{-2/(N+1)}$, $c>0$, such that the estimator G_n^* defined by (7.18) satisfies (7.22) with $\psi_n=n^{-2/(N+1)}$.

Theorems 7.3.1, 7.5.2, 7.5.3 and the lower bound for $\mathcal{G}_{\text{conv}}$ from Mammen and Tsybakov (1992) imply that the estimator (7.18) is optimal support estimator for the classes $\tilde{\mathcal{G}}$, $\mathcal{G}_{\text{conv}}$ in d_1-metric. The rates $\psi_n=n^{-\gamma/(\gamma+N-1)}$, $\psi_n=n^{-2/(N+1)}$ are the minimax rates of convergence for the classes $\tilde{\mathcal{G}}$, $\mathcal{G}_{\text{conv}}$ respectively in d_1-metric.

We should note that the MLE on ε-net given by (7.18) is mainly a theoretical construction. We introduced it to find the minimax rates of convergence. The practical support estimation procedures having optimal convergence rates could be based on the ideas of Sections 7.4 and 5.7.

CHAPTER 8. ESTIMATION OF THE DOMAIN'S AREA

8.1. PRELIMINARY DISCUSSION

In this chapter we study the problem of estimation of image functionals. We mainly deal with one example of a functional, namely, with the domain's area. Other examples are discussed in brief.

In what follows we consider the case of binary boundary fragment only, i.e. we assume that model (3.1) holds, with $f(x)=I\{x\in G\}$, and $G\in\mathcal{G}_0$. Thus, in accordance with Section 3.2.3,

$$G = \left\{ x=(x_1,x_2)\in K:\ 0\le x_1\le 1,\ 0\le x_2\le g(x_1) \right\}$$

where $g(x_1)\in\Sigma(\gamma,L,h)$, $\gamma\ge 1$. Assume that the distribution of noise ξ in observations (3.1) is Gaussian and that the design is either of type (B1) or (B2) (see Section 3.2.1).

The problem is to estimate the domain's area

$$S(G) = \text{mes } G$$

which in the case of boundary fragment can be represented as a functional

$$S(G) = S(g(\cdot)) = \int_0^1 g(x_1)dx_1 \tag{8.1}$$

defined on functions $g\in\Sigma(\gamma,L,h)$.

Let $\hat{S}_n$ be an arbitrary estimator of the area $S(G)$, i.e. a real-valued function measurable with respect to observations (3.1). The accuracy of the estimator $\hat{S}_n$ is presented by the mean squared error $E_G(\hat{S}_n-S(G))^2$.

Before we consider the domain's area estimation, let us discuss briefly the estimation of smooth functionals in nonparametric regression, since this topic is related to ours. Assume that observations in nonparametric regression model are defined by

$$Y_i = g(i/n) + \xi_i\ , \qquad i = 1,\dots,n\ , \tag{8.2}$$

with i.i.d. $(0,\sigma^2)$-Gaussian noise ξ_i, and let the function $g(x_1)$

(the regression curve) belong to the class $\Sigma(\gamma,L,h)$, $\gamma\ge 1$.

In the majority of examples related to model (8.2) smooth functionals of the curve $g(x_1)$, such as the functional (8.1), can be estimated $\sqrt{n}$-consistently, see e.g. Ibragimov and Khasminskii (1981). (There are, however, the exceptions from this rule: see Ibragimov, Nemirovskii and Khasminskii (1986), Nemirovskii (1990) for the most general results).

If one wants to estimate the functional (8.1) from observations (8.2), then the $\sqrt{n}$-consistent estimator is rather easy to find. Indeed, consider the estimator S_n^* which is the average of observations (8.2) :

$$S_n^* = (Y_1 + \ldots + Y_n)/n. \tag{8.3}$$

Since $g(x_1)$ is a smooth function the bias of S_n^* is bounded by

$$\left| E_{g(\cdot)}(S_n^*) - S(g(\cdot)) \right| = \left| (1/n)\sum_{i=1}^{n} g(i/n) - \int_0^1 g(x_1)dx_1 \right| \le L/n.$$

This implies that the normalized deviation $\sqrt{n}\left(S_n^*-S(g(\cdot))\right)$ is asymptotically zero-mean Gaussian with variance $\sigma^2=E(\xi_i^2)$. It is also easy to see that

$$\lim_{n\to\infty} E_{g(\cdot)}\left(n(S_n^*-S(g))^2 \right) = \sigma^2 \tag{8.4}$$

uniformly over $g(x_1)\in\Sigma(\gamma,L,h)$. The equality (8.4) is the definition of $\sqrt{n}$-consistency for the mean squared error.

The rate of convergence $1/\sqrt{n}$ is the minimax rate for estimation of such functionals. To prove this fact we can directly apply the results of Chapter 2 (cf. Example 2.4.1). Indeed, consider the parametric family of constant functions

$$\Theta = \left\{ g(x_1):\ g(x_1) = \theta,\ h < \theta < 1-h \right\} \subset \Sigma(\gamma,L,h).$$

It is obvious that $S(g(\cdot)) = \theta$ if $g(\cdot) \in \Theta$. If θ' and θ'' are such that $\theta''-\theta' = 1/\sqrt{n}$ then the likelihood ratio (2.11) satisfies Assumption 2.2.1. By the same argument as in Proposition 2.2.2 we immediately obtain the lower bound (2.14) with $s_n=1/\sqrt{n}$.

The rate of convergence $1/\sqrt{n}$ is the fast one. It is typical for finite-dimensional parametric problems. On the other extreme there are slow rates for estimation of "bad" functionals such as the value of a function at a fixed point (see Example 2.4.1), or the maximum of a function, etc. These rates are the same as in nonparametric regression function estimation (see Chapter 1).

Note that the substitution of optimal regression estimators of Chapter 1 does not lead to optimal estimators of S. In fact, let $g_n^* = g_n^*(x_1)$ be an optimal estimator of the regression function $g(x_1)$ based on observations (8.2), e.g. the piecewise-polynomial estimator with properly chosen bin width . As follows from the results of Section 1.6, it converges with the best possible rate $\psi_n = n^{-\gamma/(2\gamma+1)}$.Hence, the plug-in estimator of the form

$$\hat{S}_n = S(g_n^*) = \int_0^1 g_n^*(x_1)dx_1$$

has the same rate of convergence as g_n^*, which is slower than the minimax rate $1/\sqrt{n}$. (Note that for the bad functionals that we mentioned above the plug-in estimators appear to be optimal.)

Let us return to the estimation of domain's area from observations (3.1). Note that in nonparametric family of boundary fragments the best rate for a domain's estimator $\hat{G}_n$ in d_1-distance is $n^{-\gamma/(\gamma+1)}$ (see Chapter 4). This entails that the plug-in estimator $\hat{S}_n=S(\hat{G}_n)$ has the error of order $n^{-\gamma/(\gamma+1)}$. On the other hand, if G belongs to a parametric family of domains, then it is rather clear that the typical convergence rate is of order 1/n (see the next section where we discuss the parametric case).

Now, the question is, whether the minimax rate of convergence in the domain's area estimation is $n^{-\gamma/(\gamma+1)}$ or, perhaps, there exist estimators converging with the parametric rate 1/n ? The answer is rather unexpected: the minimax rate is intermediate between the typical parametric and nonparametric rates. It will be shown that up to the logarithmic factor the minimax rate is $n^{-\rho}$ where $\rho=(\gamma+1/2)/(\gamma+1)$.

8.2. DOMAIN'S AREA ESTIMATION IN CONTINUOUS PARAMETRIC MODELS

Let us show that if G belongs to a parametric family of domains then the MLE of S(G) converges with the rate 1/n. Here, as in Section 1.9, we switch to the model with continuous observations generated by Wiener fields. This is done to simplify the presentation. We mainly follow the paper of Khasminskii and Lebedev (1990). Their results concerning the domain's area estimation are quite elegant. Those readers who are not familiar with Wiener fields may pass by this section. It is useful only to keep in mind that the minimax rate of convergence is 1/n for the domain's area estimation in parametric image models.

The continuous model in this section is a two-dimensional extension of one-dimensional change-point scheme of Section 1.9. Let $W(x_1,x_2)$ be the Wiener field on the Euclidean plane, i.e. the Gaussian field with zero-mean and covariance

$$E\Big[\, W(x_1,x_2)\; W(x_1',x_2')\,\Big] = \min\,[x_1;x_1']\,\min\,[x_2;x_2'].$$

Let the observed field $Y^{(n)}(x_1,x_2) = Y(x_1,x_2)$ be of the form

$$dY(x_1,x_2) = I((x_1,x_2)\in G(\theta))dx_1dx_2 + n^{-1/2}\,dW(x_1,x_2), \tag{8.5}$$

$$(x_1,x_2) \in K = [0,1]\times[0,1].$$

Assume that the domain $G(\theta)\subseteq K$ belongs to some one-parametric family of domains

$$\mathcal{G}_\Theta = \Big\{\, G = G(\theta),\ \theta\in\Theta \subset R^1 \,\Big\}.$$

Surely, some regularity conditions should be imposed on the set of domains $\mathcal{G}_\Theta$. Before we introduce such conditions let us discuss the simplest case (the boundary fragment). Assume that the edge $\partial G(\theta)$ of $G(\theta)$ is

$$\partial G(\theta) = \Big\{\, (x_1,x_2)\in K:\ x_2 = g(x_1,\theta),\ 0\le x_1\le 1 \,\Big\},\ \theta\in\Theta.$$

If the function $g(x_1,\theta)$ is smooth then

$$(8.6)\qquad \partial G(\theta+u)=\left\{(x_1,x_2)\in K:\ x_2=g(x_1,\theta)+u\partial g(x_1,\theta)/\partial\theta+o(u),\ 0\le x_1\le 1\right\}$$

where $o(u)/u \to 0$ as $u \to 0$ uniformly on $0\le x_1\le 1$. For any fixed real values u and v define

$$G_n(u,v|\theta) = \left(G(\theta+u/n)\ \Delta\ G(\theta)\right) \cap \left(G(\theta+v/n)\ \Delta\ G(\theta)\right).$$

Denote the area of $G(\theta)$ by $S(G(\theta))$. If one takes into account the first terms of the Taylor expansion in (8.6) it is easy to see that

$$(8.7)\qquad \lim_{n\to\infty}\left(n\ S(G_n(u,v|\theta))\right) = \kappa(u,v)\int_0^1 |\partial g(x_1,\theta)/\partial\theta|\ dx_1$$

where

$$\kappa(u,v) = \begin{cases} \min[u;v] & \text{if } u>0,\ v>0,\\ 0 & \text{if } \operatorname{sign}(u)\ne\operatorname{sign}(v),\\ \min[-u;-v] & \text{if } u<0,\ v<0. \end{cases}$$

In the general case of domain $G(\theta)\in\mathcal{G}_\Theta$ we take (8.7) as definition of regularity.

8.2.1. ASSUMPTION. For any $\theta\in\Theta$ and arbitrary fixed real values u and v the equality holds

$$\lim_{n\to\infty}\left(n\ S(G_n(u,v|\theta))\right) = C(\theta)\ \kappa(u,v)$$

where $C(\theta)$ is a continuous and strictly positive function on Θ.

The consequence of this assumption is the following:

$$(8.8)\qquad n\ S(G(\theta+u)\ \Delta\ G(\theta)) = |u|C(\theta) + o(1)$$

as $u \to 0$. Indeed, $\kappa(u,u) = |u|$, and thus (8.8) follows.

8.2.2. PROPOSITION. If Assumption 8.2.1 holds, then the normalized log-likelihood ratio

$$Z_{\theta}^{(n)}(u) = \log\left(dP_{\theta+u/n} \;/\; dP_{\theta}\right)(Y^{(n)}(\cdot))$$

converges as $n \to \infty$ weakly with respect to P_{θ}-probability to the stochastic process

$$Z_0(C(\theta)u) = W_0(C(\theta)u) - C(\theta)|u|/2, \; u \in R^1,$$

where $W_0(u)$ is the two-sided Wiener process defined in Section 1.9.

Proof. The relation

$$(8.9) \qquad Z_{\theta}^{(n)}(u) =$$

$$\sqrt{n}\int_0^1\int_0^1\Big[I\{(x_1,x_2)\in G(\theta+u/n)\}-I\{(x_1,x_2)\in G(\theta)\}\Big]dW(x_1,x_2) -$$

$$-\frac{1}{2}n\int_0^1\int_0^1\Big[I\{(x_1,x_2)\in G(\theta+u/n)\}-I\{(x_1,x_2)\in G(\theta)\}\Big]^2 dx_1dx_2$$

is true for the Wiener fields. Expectation of $Z_{\theta}^{(n)}(u)$ is defined by the second summand in the right-hand side of (8.9):

$$E\Big[Z_{\theta}^{(n)}(u)\Big] = -\frac{n}{2}\int_0^1\int_0^1\Big[I\{(x_1,x_2)\in G(\theta+u/n)\}-I\{(x_1,x_2)\in G(\theta)\}\Big]^2 dx_1dx_2 =$$

$$= -\frac{1}{2}nS(G(\theta+u/n)\Delta G(\theta)) = -\frac{1}{2}|u|C(\theta) + o(1)$$

in accordance with (8.8). Hence

$$(8.10) \qquad \lim_{n\to\infty} E\Big[Z_{\theta}^{(n)}(u)\Big] = -\frac{1}{2}C(\theta)|u|.$$

Further, the properties of stochastic integrals with respect to the Wiener field imply that the first summand in the right-hand side of (8.9) has zero-mean, and its covariance is

$$(8.11)\qquad E\Big[Z_\theta^{(n)}(u)Z_\theta^{(n)}(v)\Big] = n\int_0^1\int_0^1\Big(\Big[I\{(x_1,x_2)\in G(\theta+u/n)\}-$$

$$-I\{(x_1,x_2)\in G(\theta)\}\Big]\Big[I\{(x_1,x_2)\in G(\theta+v/n)\}-I\{(x_1,x_2)\in G(\theta)\}\Big]dx_1dx_2 =$$

$$= n\ \mathrm{mes}\Big((G(\theta+u/n)\ \Delta\ G(\theta))\cap(G(\theta+v/n)\ \Delta\ G(\theta))\Big) =$$

$$= n\ S(G_n(u,v|\theta)) = \kappa(u,v)C(\theta) + o(1)$$

as $n\to\infty$. Since $Z_\theta^{(n)}(u)$ is a Gaussian stochastic process it follows from (8.10) and (8.11) that its limit distribution coincides with $Z_0(C(\theta)u)$. □

8.2.3. COROLLARY. Denote by u_0^* the maximizer of $Z_0(u)$, and let θ_0^* be the maximum likelihood estimator (MLE) of θ based on observations (8.5). If for all $\theta_1\neq\theta_2$, $\theta_1,\theta_2\in\Theta$, the following relation holds:

$$(8.12)\qquad d_1(G(\theta_1),G(\theta_2)) > 0$$

then the normalized deviation $nC(\theta)(\theta_0^*-\theta)$ converges weakly to the distribution of u_0^* (cf. Section 1.9). The moments of $nC(\theta)(\theta_0^*-\theta)$ converge to the corresponding moments of u_0^*. Note that the bigger $C(\theta)$ the faster θ_0^* converges to θ.

Now we start with the special case when parameter θ is the domain's area.

8.2.4. ASSUMPTION.

(i) $\theta = S(G(\theta))$ for all $G(\theta)\in\mathcal{G}_\Theta$.

(ii) The family $\mathcal{G}_\Theta$ is an <u>expanding family</u> in θ, i.e. the inequality $\theta_1<\theta_2$ implies $G(\theta_1)\subset G(\theta_2)$.

8.2.5. PROPOSITION. If Assumption 8.2.4 holds, then $C(\theta)\equiv 1$.

Proof. As shown in the proof of Proposition 8.2.2,

$$E\left[Z_\theta^{(n)}(u)\right] = -\frac{1}{2}nS(G(\theta+u/n)\Delta G(\theta)).$$

Since $\mathcal{G}_\theta$ is an expanding family and parameter θ coincides with the area $S(G(\theta))$

$$S(G(\theta+u/n)\Delta G(\theta)) = \begin{cases} S(G(\theta+u/n)) - S(\theta) & \text{if } u>0, \\ S(\theta) - S(G(\theta+u/n)) & \text{if } u<0 . \end{cases}$$

Thus, the exact equality holds

$$E\left[Z_\theta^{(n)}(u)\right] = -\frac{1}{2}|u|.$$

Similar argument proves (8.11) with $C(\theta)\equiv 1$ and $o(1)\equiv 0$. □

8.2.6. PROPOSITION. Let inequality (8.12) and Assumptions 8.2.4 (i), 8.2.1 hold. Then the squared risk of the MLE θ_0^* has the following upper bound for all $\theta\in\Theta$:

$$\lim_{n\to\infty} E_\theta\left[n^2(\theta_0^*-\theta)^2\right] \le E(u_0^*)^2$$

Proof. Taking into account Corollary 8.2.3 we note that it is sufficient to prove the inequality $C(\theta)\ge 1$. Using the identity $\theta=S(G(\theta))$ one can write

$$E\left[Z_\theta^{(n)}(u)\right] = -\frac{1}{2}n\int_0^1\int_0^1\left[I\{(x_1,x_2)\in G(\theta+u/n)\}-I\{(x_1,x_2)\in G(\theta)\}\right]^2 dx_1dx_2 =$$

$$=-\frac{1}{2}n\left[\int_0^1\int_0^1 I^2\{(x_1,x_2)\in G(\theta+u/n)\}dx_1dx_2 +\right.$$

$$+\int_0^1\int_0^1 I^2\{(x_1,x_2)\in G(\theta)\}dx_1dx_2 -$$

$$- 2\int_0^1\int_0^1 I\{(x_1,x_2)\in G(\theta+u/n)\}\ I\{(x_1,x_2)\in G(\theta)\}dx_1dx_2 \Big] =$$

$$= -\frac{1}{2}n\Big(S(G(\theta+u/n)) + S(G(\theta)) - 2\ S(G(\theta)\cap(G(\theta+u/n))\Big) \le$$

$$\le -\frac{1}{2}n\Big(\ \theta + u/n + \theta - 2\ \min[\theta+u/n;\theta]\ \Big) = -\frac{1}{2}|u|.$$

Comparing the left-hand side with (8.10) we prove the proposition. □

Let us draw some conclusions. In parametric setup of the domain's area estimation problem the convergence rate of MLE is 1/n, as for the one-dimensional change-point problem. Propositions 8.2.5 and 8.2.6 show that the case of expanding families is the most difficult for estimation since for this case the constant $C(\theta)$ has the least possible value: $C(\theta)=1$.

8.3. THEOREM ON THE LOWER BOUND

Let us return to the discrete model with observations (3.1). We start with the lower bound for the domain's area estimation.

8.3.1. THEOREM. Let ξ_i in (3.1) be zero-mean Gaussian random variables and let the design satisfy (B2). Then for any estimator $\hat{S}_n$ of the domain's area S(G) inequality holds

$$\liminf_{n\to\infty}\ \sup_{G\in\mathcal{G}_0}\ E_G(n^{2\rho}(\hat{S}_n-S(G))^2) \ge C_0 \tag{8.13}$$

with some positive C_0 and $\rho=(\gamma+1/2)/(\gamma+1)$.

Proof. We shall obtain the lower bound (8.13) comparing the minimax and the Bayesian risks for specially defined probabilistic measure on $\mathcal{G}_0$. This method is standard and it has already been used in the proof of Assouad's lemma in Section 2.6. Introduce a parametric family of domains $G(\theta)$, $\theta=(\theta_1,\dots,\theta_M)$, $\theta_\ell\in\{0,1\}$, defined in the following way:

$$G(\theta) = \bigcup_{\ell=1}^{M} G_\ell(\theta_\ell)$$

where

$$G_\ell(\theta_\ell) = \{ (x_1,x_2)\in K:\ 0 \le x_2 \le 1/2+\theta_\ell(g_{\ell n}(x_1)-1/2)\},\ \ell=1,\dots,M,$$

with $\delta_n = n^{-1/(\gamma+1)}$, $M = 1/\delta_n$ and $g_{\ell n}$ as in Section 3.3.4. Remind that the function $g_{\ell n}(x_1)$ belongs to the class $\Sigma(\gamma,L)$; $g_{\ell n}(x_1)\ge 1/2$, and $g_{\ell n}(x_1)=1/2$ if $x_1\notin[u_{\ell-1},u_\ell)$,

$$\int g_{\ell n}(x_1)\,dx_1=\frac{1}{2}+c\delta_n^{\gamma+1}.$$

Thus $\{G(\theta)\}\subset\mathcal{G}_0$. Now, the minimax risk is bounded from below by the Bayesian risk for $\{G(\theta)\}$ provided that θ_ℓ are independent random variables taking values 0 and 1 with probability 1/2:

$$(8.14)\qquad \inf_{\hat S_n}\sup_{G\in\mathcal{G}_0} E_G(n^{2\rho}(\hat S_n - S(G))^2) \ge$$

$$\ge \inf_{\hat S_n} E_\theta E_{G(\theta)}\left(n^{2\rho}(\hat S_n - S(G(\theta)))^2\right) =$$

$$= c\inf_{\hat S_n} E_\theta E_{G(\theta)}\left((\tilde S_n - M^{-1/2}\sum_{\ell=1}^{M}\theta_\ell)^2\right).$$

In (8.14), as usual, $E_{G(\theta)}$ denotes the conditional expectation with respect to the distribution of $((X_1,Y_1),\dots,(X_n,Y_n))$ for fixed θ, while E_θ is the expectation with respect to a priori distribution of $\theta=(\theta_1,\dots,\theta_M)$. The last identity in (8.14) is due to the fact that

$$\hat S_n-S(G(\theta)) = \hat S_n-\left(\frac{1}{2} + cn^{-1}\sum_{\ell=1}^{M}\theta_\ell\right) = cn^{-\rho}\left(\tilde S_n - M^{-1/2}\sum_{\ell=1}^{M}\theta_\ell\right),$$

and the mapping $\tilde S_n=(\hat S_n-\frac{1}{2})n^\rho/c$ transforms the set of all possible estimates into itself. The last expression in (8.14) does not exceed $cE_{\theta}E_{G(\theta)}\left((\hat{\phi}_n - M^{-1/2}\sum_{\ell=1}^{M}\theta_{\ell})^2\right)$ where $\hat{\phi}_n$ is the Bayesian estimate of $M^{-1/2}\sum_{\ell=1}^{M}\theta_{\ell}$. The Bayesian estimate $\hat{\theta}_{\ell}$ of a single parameter θ_{ℓ} is equal to a posteriori probability:

$$\hat{\theta}_{\ell} = P(\theta_{\ell}=1|X^{(n)}) = P_{G_{\ell}}(X^{(n)})/\left(P_{G_{\ell}}(X^{(n)})+P_{G_0}(X^{(n)})\right).$$

Note that $\hat{\theta}_{\ell}$ depends only on those observations (X_i,Y_i) for which the X_i's belong to $G_{\ell}\backslash G_0$. This entails that $\hat{\theta}_{\ell}$'s are independent for different ℓ. Since also $\hat{\theta}_{\ell}$ is a Bayesian estimator, we have $E_{\theta}(\hat{\theta}_{\ell}-\theta_{\ell})=0$. Clearly, $\hat{\phi}_n = M^{-1/2}\sum_{\ell=1}^{M}\hat{\theta}_{\ell}$. Thus, one obtains

$$E_{\theta}E_{G(\theta)}\left((\hat{\phi}_n - M^{-1/2}\sum_{\ell=1}^{M}\theta_{\ell})^2\right) = E_{\theta_1}E_{G_1(\theta_1)}\left((\hat{\theta}_1-\theta_1)^2\right),$$

and (3.11) implies that the last expectation is bounded from below by some positive constant. This proves the theorem. □

8.4. OPTIMAL ESTIMATOR FOR THE DOMAIN'S AREA

8.4.1. THEOREM. Under the conditions of Theorem 8.3.1 there exists an estimator S_n^* such that

$$\limsup_{n\to\infty}\ \sup_{G\in\mathcal{G}_0} E_G\left(n^{2\rho}(\log n)^{-\gamma/(\gamma+1)}(S_n^*-S(G))^2\right) \le C_1 \tag{8.15}$$

with some constant C_1 and $\rho=(\gamma+1/2)/(\gamma+1)$.

Neglecting the log-factor appearing in (8.15) we call S_n^* <u>optimal estimator</u>. The explicit form of S_n^* is given below.

8.4.2. REMARK. It seems probable that the log-factor which spoils the rate of convergence in (8.15) may disappear for some other estimators S_n^*. However, we don't have the proof of this. The problem of finding exact minimax rate remains open.

8.4.3. HEURISTIC ARGUMENT. The following argument suggests the way of constructing an estimator satisfying (8.15) and explains the improvement in minimax rate for the area estimation as compared with edge estimation.

Suppose that an estimator $g_*(x_1)$ is given independent of observations (3.1) and such that

$$\sup_{0\le x_1\le 1} |g_*(x_1)-g(x_1)| \le C_*\psi_n \tag{8.16}$$

with some positive C_* and $\psi_n=((\log n)/n)^{\gamma/(\gamma+1)}$.

Define the curved strip U_0 of width $2C_*\psi_n$ around the function $g_*(x_1)$:

$$U_0=\left\{x=(x_1,x_2)\in K:\ |x_2-g_*(x_1)| \le C_*\psi_n\right\}.$$

It suffices to estimate the area of $G\cap U_0$ since

$$S(G) = S(G\cap\bar{U}_0) + S(G\cap U_0) = \int_0^1 (g_*(x_1)-C_*\psi_n)dx_1 + S(G\cap U_0).$$

The area $S(G\cap U_0)$ can be estimated by

$$\hat{S}_n = n^{-1}\sum_i I\{X_i\in U_0\}Y_i. \tag{8.17}$$

The estimator $\hat{S}_n$ is unbiased,

$$E_G(\hat{S}_n) = S(G\cap U_0),$$

under the designs (B1) and (B2), and its variance is $O(\psi_n/n)$. Thus uniformly over $G\in\mathcal{G}_0$

$$E_G\left((n/\psi_n)(\hat{S}_n-S(G\cap U_0))^2\right) \le c_1$$

for a positive c_1. But n/ψ_n is equivalent to $n^{2\rho}$ up to the log-factor.

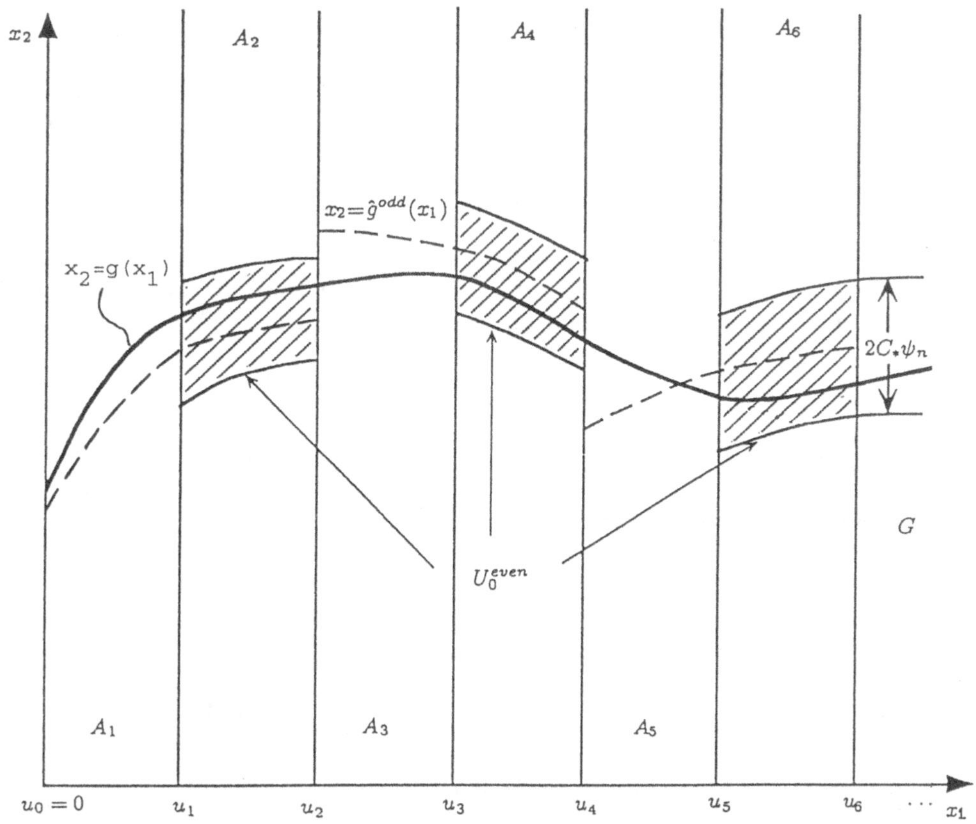

Fig. 8.1. To the definition of optimal area estimator.

8.4.4. OPTIMAL ESTIMATOR. To use the heuristic idea of Section 8.4.3 one has to define an estimator $g_*(x_1)$ satisfying (8.16) in some probabilistic sense. We consider here the case of design (B2).

Let A_ℓ be rectangular slices introduced in Section 4.1, with the width $\delta_n=((\log n)/n)^{1/(\gamma+1)}$, and let G^{odd} and G^{even} be the subsets of G corresponding to odd and even slice indices (see Remark 4.3.3). It suffices to estimate $S(G^{odd})$ and $S(G^{even})$ separately.

Consider the estimate of $S(G^{even})$. Let $\hat{g}^{odd}(x_1)$ be the edge estimator under the zebra design ($B2^{odd}$) as defined in Remark 4.3.3 (see Fig. 8.1). Denote

$$U_0^{even} = \left\{ x=(x_1,x_2)\in K^{even}: |x_2-\hat{g}^{odd}(x_1)| \le C_*\psi_n \right\}$$

where $\psi_n=((\log n)/n)^{\gamma/(\gamma+1)}$, K^{even} is the union of even slices $A_\ell \cap K$, and C_* is large enough so that the inequality

$$\sup_{0\le x_1\le 1} |\hat{g}^{odd}(x_1)-g(x_1)| \le C_*\psi_n/2 \tag{8.18}$$

holds with P_G-probability at least $1-n^{-2}$. Such a C_* can be chosen in accordance with (4.15). Following the argument of Section 8.4.3 define the estimator of $S(G^{even})$ as

$$\hat{S}_n^{even} = \sum_{\ell \text{ even}} \int_{u_{\ell-1}}^{u_\ell} (\hat{g}^{odd}(x_1)-C_*\psi_n)dx_1 + n^{-1}\sum_i I\{X_i\in U_0^{even}\}Y_i . \tag{8.19}$$

Converting the odd and even slices one gets a similar estimator $\hat{S}_n^{odd}$ for $S(G^{odd})$. Finally, define $S_n^* = \hat{S}_n^{even} + \hat{S}_n^{odd}$.

Proof of Theorem 8.4.1. Note that

$$(S_n^*-S(G))^2 \le 2(\hat{S}_n^{even}-S(G^{even}))^2 + 2(\hat{S}_n^{odd}-S(G^{odd}))^2 .$$

Consider for definiteness the first term on the right-hand side. Define the random event

$$\mathcal{A}_n = \left\{ \sup_{0\le x_1\le 1} |\hat{g}^{odd}(x_1)-g(x_1)| \le C_*\psi_n/2 \right\} .$$

If $\mathcal{A}_n$ holds, then

$$S(G^{even}) = \sum_{\ell \text{ even}} \int_{u_{\ell-1}}^{u_\ell} (\hat{g}^{odd}(x_1) - C_* \psi_n)\, dx_1 + S(G \cap U_0^{even}). \tag{8.20}$$

Equalities (8.19) and (8.20) entail

$$\left(\hat{S}_n^{even} - S(G^{even})\right) I(\mathcal{A}_n) = \tilde{\eta}_n I(\mathcal{A}_n)$$

where $\tilde{\eta}_n = n^{-1} \sum_i I\{X_i \in U_0^{even}\} Y_i - S(G \cap U_0^{even})$. Note that $\tilde{\eta}_n$ and $I(\mathcal{A}_n)$ are independent, and $E_G[\tilde{\eta}_n] = 0$; $E_G[\tilde{\eta}_n^{\,2}] \le c_2(\psi_n/n)$ with some constant $c_2 > 0$. Thus, we have

$$E_G\left[n^{2\rho}(\log n)^{-\gamma/(\gamma+1)}\left(\hat{S}_n^{even} - S(G^{even})\right)^2\right] = (n/\psi_n) E_G\left[\ \tilde{\eta}_n^2\ I(\mathcal{A}_n)\right] +$$

$$+ (n/\psi_n) E_G\left[\left(\hat{S}_n^{even} - S(G^{even})\ \right)^2 I(\bar{\mathcal{A}}_n)\right] \le c_2 + c_3 (n/\psi_n) P_G(\bar{\mathcal{A}}_n).$$

Due to the choice of C_* in (8.18) the last probability is bounded: $P_G(\bar{\mathcal{A}}_n) \le n^{-2}$, so that $(n/\psi_n)\ n^{-2} = (n\ \psi_n)^{-1} \le 1$ if n is large enough. Hence , the last sum is bounded. □

8.5. GENERALIZATIONS AND EXTENSIONS

Let us discuss some generalizations of the results of Sections 8.3 and 8.4. First, we note that these results can be extended to multidimensional images.

8.5.1. MULTIDIMENSIONAL CASE. Assume that the domain G belongs to the class $\mathcal{G}_{0,N}$ of multidimensional boundary fragments, $N \ge 2$. Let $(B2_N)$ be the natural extension of the design (B2) to N-dimensions, $N \ge 2$, and let $(B2_N^{odd})$, $(B2_N^{even})$ be the N-dimensional versions of zebra designs defined in Remark 4.3.3 (we assume now that A_ℓ are N-dimensional slices as in Section 5.1).

Consider the edge estimator $\hat{g}(x_1, \ldots, x_{N-1})$ of Section 5.1 with $\psi_n = ((\log n)/n)^{\gamma/(\gamma+N-1)}$ under the N-dimensional zebra design $(B2_N^{odd})$. Denote this estimator $\hat{g}^{odd}(x_1, \ldots, x_{N-1})$. Define

$$U_0^{\text{even}} = \left\{ x=(x_1,\ldots,x_N)\in K^{\text{even}}:\ |x_N-\hat{g}^{\text{odd}}(x_1,\ldots,x_{N-1})| \le C_*\psi_n \right\}$$

where $C_*>0$ is large enough so that inequality

$$(8.21)\qquad \sup_{0\le x_1,\ldots,x_{N-1}\le 1} |\hat{g}^{\text{odd}}(x_1,\ldots,x_{N-1})-g(x_1,\ldots,x_{N-1})| \le C_*\psi_n/2$$

holds with P_G-probability at least $1-n^{-2}$. The existence of such C_* is shown as in two-dimensional case. It appears in the proof of Theorem 5.1.6 (ii) (however, we omitted this proof in Chapter 5 because of its similarity to the proof of Theorem 4.4.1).

Define the estimator

$$\hat{S}_n^{\text{even}} = \sum_{\ell\ \text{even}} \int_{Q_\ell} (\hat{g}^{\text{odd}}(x_1,\ldots,x_{N-1})-C_*\psi_n)\,dx_1\ldots dx_{N-1} +$$

$$+\ n^{-1}\sum_i I\{X_i\in U_0^{\text{even}}\}Y_i$$

($\hat{S}_n^{\text{odd}}$ is defined similarly). Here Q_ℓ's are the cubes forming a partition of $[0,1]^{N-1}$ (see Section 5.1). Set $S_n^* = \hat{S}_n^{\text{odd}} + \hat{S}_n^{\text{even}}$.

The following is proved exactly in the same way as Theorems 8.4.1 and 8.3.1:

8.5.2. THEOREM. Let ξ_i in (3.1) be zero-mean Gaussian random variables and let the design be of type $(B2_N)$. Then the estimator S_n^* defined in Section 8.5.1 satisfies

$$(8.22)\qquad \limsup_{n\to\infty}\ \sup_{G\in\mathcal{G}_{0,N}} E_G\left(n^{2\rho}(\log n)^{-\gamma/(\gamma+N-1)}(S_n^*-S(G))^2\right) \le C_1$$

with some constant C_1 and $\rho=(\gamma+(N-1)/2)/(\gamma+N-1)$. The rate $n^{-\rho}$ is optimal up to the logarithmic factor, i.e. for any estimator $\hat{S}_n$ of the domain's area $S(G)$ the following inequality holds

$$\liminf_{n\to\infty}\ \sup_{G\in\mathcal{G}_{0,N}} E_G(n^{2\rho}(\hat{S}_n-S(G))^2) \ge C_0$$

with some positive C_0.

8.5.3. REMARK. (i) Theorems 8.3.1 and 8.4.1 remain valid under the design (B1). The definition of S_n^* is slightly changed then. In particular, $\hat{g}^{odd}$ should be the estimator based on the subsample containing all pairs (X_i, Y_i) with odd indices i. The estimator S_n^* is then given by

$$(8.23) \qquad S_n^* = \int_K I\{x_N \le \hat{g}^{odd}(x_1,\ldots,x_{N-1}) - C_*\psi_n\}dx + $$

$$+ \frac{2}{n} \sum_{i \text{ even}} I\{X_i \in \tilde{U}_0\} Y_i$$

where

$$(8.24) \qquad \tilde{U}_0 = \left\{ x=(x_1,\ldots,x_N) \in K:\ |x_N - \hat{g}^{odd}(x_1,\ldots,x_{N-1})| \le C_*\psi_n \right\}.$$

Here and later we assume without loss of generality that the number n is even (otherwise the second summand in the expression for S_n^* should be replaced by $\mathrm{card}\{X_i:\ i \text{ even}\}^{-1} \sum_{i \text{ even}} I\{X_i \in \tilde{U}_0\} Y_i$).

(ii) The assumption of additive noise model (3.1) with Gaussian ξ_i's may be replaced by Assumption 5.1.5 in what concerns the upper bounds of Theorems 8.4.1 and 8.5.2. For example, the multiplicative model (5.1) satisfies this assumption.

8.5.4. ESTIMATION OF DOMAIN'S FUNCTIONALS. The technique of Sections 8.3 and 8.4 applies to estimation of domain's functionals of the form

$$(8.25) \qquad S(G) = \int_G \varphi(x)\ dx$$

where $\varphi(x)$ is a given function on $K=[0,1]^N$. If $\varphi(x) \equiv 1$ then S(G) is the domain's area.

There are more general functionals close to (8.25):

$$S(G) = \Phi\left(\int_G \varphi_1(x)\ dx, \ldots, \int_G \varphi_s(x)\ dx \right) \tag{8.26}$$

where $\Phi(u_1,\ldots,u_s)$ is a smooth function of the real-valued arguments $u_1,\ldots,u_s$. An important example of the general functional (8.26) is <u>the center</u> x^* of G (cf. Section 5.7):

$$S(G) = x^* = (\text{mes } G)^{-1} \int_G x\ dx$$

Note that x^* is a vector composed of N functionals of the form (8.26) with $s=2$, $\Phi(u_1,u_2)=u_2/u_1$, and $\varphi_1(x)=1$, $\varphi_2(x)=x_j$, $j=1,\ldots,N$. Another example is the "spread" of G:

$$S(G) = (\text{mes } G)^{-1} \int_G |x-x^*|^2\ dx.$$

The smooth functionals (8.26) can be estimated in an asymptotically optimal way if one combines the method of this section with the idea of smooth regression functionals estimation (see Ibragimov and Khasminskii (1981), Ch.7). To describe the latter in brief assume that some preliminary sufficiently accurate estimators $\hat{u}_1,\ldots,\hat{u}_s$ of the functionals $\int_G \varphi_1(x)dx, \ldots, \int_G \varphi_s(x)dx$ are given. As a rule, $\Phi(\hat{u}_1,\ldots,\hat{u}_s)$ is not the optimal estimator for (8.26). Using the Taylor expansion, one gets

$$S(G) \approx \Phi(\hat{u}_1,\ldots,\hat{u}_s) + \sum_{j=1}^{s} \partial\Phi(\hat{u}_1,\ldots,\hat{u}_s)/\partial u_j \left(\int_G \varphi_j(x)dx - \hat{u}_j \right).$$

Now, to obtain an asymptotically optimal estimator of (8.26) it suffices to estimate $\int_G \varphi_j(x)dx$ independently of $\hat{u}_1,\ldots,\hat{u}_s$. This can be done by a routine splitting of the sample into independent parts.

Here we consider only the estimation problem for the simplest functional (8.25). Assume that the design is of type

(B1). Introduce the following estimator of this functional:

$$(8.27)\qquad S_n^* = \int_K \varphi(x) I\{x_N \le \hat{g}^{odd}(x_1,\dots,x_{N-1}) - C_*\psi_n\}dx + $$

$$+ \frac{2}{n} \sum_{i \text{ even}} I\{X_i \in \tilde{U}_0\} Y_i \varphi(X_i)$$

where $\hat{g}^{odd}$ and $\tilde{U}_0$ are as in Remark 8.5.3.

8.5.5. THEOREM. Let ξ_i in (3.1) be zero-mean Gaussian random variables and let the design be of type (B1), $N \ge 2$. If $\varphi(x)$ is a bounded Borel function then the estimator (8.27) satisfies (8.22).

The proof, again, follows the same lines as in Theorem 8.4.1, and we leave it to the reader. Also, the lower bound is valid as in Theorem 8.3.1 for arbitrary estimators of (8.25).

8.6. FUNCTIONALS OF SUPPORT OF A DENSITY

Another extension of the results of this chapter is related to support estimation problem discussed in Chapter 7. Let $X_1,\dots,X_n$ be i.i.d. random variables uniformly distributed in $G \subset K = [0,1]^N$. The problem is to estimate the functional (8.25) where φ is a bounded Borel function. Assume that the support G belongs to the class of boundary fragments $\mathcal{G}_{0,N}$. To start with, we define the estimator analogous to (8.27):

$$(8.28)\qquad S_n^* = \int_K \varphi(x) I\{x_N \le \hat{g}^{odd}(x_1,\dots,x_{N-1}) - C_*\psi_n\}dx + $$

$$+ \frac{2}{n} \left(\text{mes } G_n^{odd}\right) \sum_{i \text{ even}} I\{X_i \in \tilde{U}_0\} \varphi(X_i).$$

Here $\hat{g}^{odd}(x_1,\dots,x_{N-1})$ is the optimal edge estimator based on the odd subsample, i. e. a solution of (7.16) with $M_n = (n/(\log n))^{1/(\gamma+N-1)}$; $\tilde{U}_0$ is defined as in (8.24), $\psi_n = ((\log n)/n)^{\gamma/(\gamma+N-1)}$, G_n^{odd} is the closure of the set

$$\{0 \le x_1, \dots, x_{N-1} \le 1,\ x_N \le \hat{g}^{odd}(x_1, \dots, x_{N-1})\}.$$

(Warning: the notation in this section is the same as earlier in this chapter, though the statistical model is different!) The constant C_* is chosen in such a way that (8.21) holds with P_G-probability at least $1-n^{-2}$. Such choice of C_* is possible: for N=2 it follows from the proof of Theorem 7.4.1 (see (7.15)); the case N>2 is treated quite similarly.

8.6.1. THEOREM. Assume that $X_1, \dots, X_n$ are i.i.d. random variables uniformly distributed in $G \in \mathcal{G}_{0,N}$. Let S(G) be the functional (8.25) where φ is a bounded Borel function. Then for the estimator (8.28) we have

$$\sup_{G \in \mathcal{G}_{0,N}} E_G\left((S_n^* - S(G))^2\right) = O\left(\max\{\psi_n/n; \psi_n^4\}\right) = \tag{8.29}$$

$$= O\left(\max\{n^{-2\rho}(\log n)^{\gamma/(\gamma+N-1)}, (\log n/n)^{-4\gamma/(\gamma+N-1)}\}\right),\quad n\to\infty,$$

where $\rho = (\gamma + (N-1)/2)/(\gamma+N-1)$.

<u>Proof</u>. Denote by $\mathcal{A}_n^N$ the random event given in (8.21). If $\mathcal{A}_n^N$ holds then $G_n^{odd} \subset G$. Thus, we have

$$(S_n^* - S(G))I(\mathcal{A}_n^N) = (S_{n0} + S_{n1})I(\mathcal{A}_n^N)$$

where

$$S_{n0} = \frac{2}{n} \sum_{i\ \text{even}} \varphi_i,$$

$$\varphi_i = I\{X_i \notin G_n^{odd}\}\varphi(X_i) \text{ mes } G -$$

$$- \int_K \varphi(x) I\{x \in G \setminus G_n^{odd}\} dx,$$

$$S_{n1} = (\text{mes } G_n^{\text{odd}} - \text{mes } G) S_{n2},$$

$$S_{n2} = \frac{2}{n} \sum_{i \text{ even}} I\{X_i \notin G_n^{\text{odd}}\} \varphi(X_i).$$

Clearly, φ_i, i even, are i.i.d. random variables, and

$$E_G(\varphi_i I(\mathscr{A}_n^N) \mid \{X_i, \text{ i odd}\}) = 0,$$

$$E_G(\varphi_i^2 I(\mathscr{A}_n^N) \mid \{X_i, \text{ i odd}\}) \le$$

$$\le 2I(\mathscr{A}_n^N) \Big[\text{mes } G \int_K \varphi^2(x) I\{x \in G \setminus G_n^{\text{odd}}\} dx +$$

$$+ \Big(\int_K \varphi(x) I\{x \in G \setminus G_n^{\text{odd}}\} dx \Big)^2 \Big] \le$$

$$\le c_4 \text{ mes}(G \setminus G_n^{\text{odd}}) I(\mathscr{A}_n^N)$$

where $c_4 > 0$ is a constant.

Note that $\mathscr{A}_n^N$ depends only on X_i with i odd. Hence

$$E_G(S_{n0}^2 I(\mathscr{A}_n^N)) \le c_4 n^{-1} E_G(\text{mes}(G \setminus G_n^{\text{odd}}) I(\mathscr{A}_n^N)). \tag{8.30}$$

If $\mathscr{A}_n^N$ holds then

$$\text{mes}(G \setminus G_n^{\text{odd}}) \le c_5 \psi_n. \tag{8.31}$$

where $c_5 > 0$ is a constant that does not depend on G. It follows from (8.30), (8.31) that uniformly in $G \in \mathscr{G}_{0,N}$

(8.32) $$E_G(S_{n0}^2 I(\mathcal{A}_n^N)) = O(\psi_n/n), \quad n\to\infty.$$

Now,

(8.33) $$E_G(S_{n2}^2 I(\mathcal{A}_n^N) \mid \{X_i,\ i \text{ odd}\}) =$$

$$=\left(\frac{2}{n}\right)^2 \sum_{i,j \text{ even}} E_G\Big(I\{X_i \notin G_n^{odd}\} I\{X_j \notin G_n^{odd}\} \varphi(X_i)\varphi(X_j) I(\mathcal{A}_n^N) \mid \{X_i, i \text{ odd}\}\Big) =$$

$$= \left[(1 + O(1/n))\left((\text{mes } G)^{-1}\int_K \varphi(x) I\{x\in G\setminus G_n^{odd}\}dx\right)^2 + \right.$$

$$\left. + O(1/n)(\text{mes } G)^{-1}\int_K \varphi^2(x) I\{x\in G\setminus G_n^{odd}\}dx\right] I(\mathcal{A}_n^N) \le$$

$$\le c_6\left[\text{mes}^2(G\setminus G_n^{odd}) + (1/n)\text{mes}(G\setminus G_n^{odd})\right] I(\mathcal{A}_n^N) = O(\psi_n^2), n\to\infty.$$

Hence

(8.34) $$E_G(S_{n1}^2 I(\mathcal{A}_n^N)) =$$

$$= E_G((\text{mes } G_n^{*odd} - \text{mes } G)^2 I(\mathcal{A}_n^N)) E_G(S_{n2}^2 I(\mathcal{A}_n^N) \mid \{X_i, i \text{ odd}\}) =$$

$$= O(\psi_n^4), \quad n\to\infty.$$

Finally, we note that

(8.35) $$P_G(\bar{\mathcal{A}}_n^N) \le n^{-2}.$$

Now,

$$E_G\left((S_n^* - S(G))^2\right) \le P_G(\bar{\mathcal{A}}_n^N) + 2\left[E_G(S_{n0}^2 I(\mathcal{A}_n^N)) + E_G(S_{n1}^2 I(\mathcal{A}_n^N))\right],$$

and the theorem follows from (8.32), (8.34)-(8.35). □

Note that the right-hand side of (8.29) is the maximum of two terms: the usual one, $O(\psi_n/n)$ (as in Theorem 8.4.1), and the new one, $O(\psi_n^4)$. A brief analysis of the proof shows that if we could substitute mes G instead of mes G_n^{odd} into the second summand in (8.28), then only the term $O(\psi_n/n)$ would be present in (8.29), as in Theorem 8.4.1. But since the value of mes G is not available, the error term $O(\psi_n^4)$ related to the estimation of mes G comes in (see (8.34)). This term is negligible with respect to ψ_n/n if the inequality holds: $N-1<2\gamma$. In particular, this is true for $N=2$, $\gamma\geq 1$, and we conclude that in this case our estimator is optimal up to the logarithmic factor. But for N large this inequality is violated, and an additional effort is needed to construct the optimal estimator.

The idea of improvement is quite simple: assume that $\hat{\mu}_{1,n}$ is an estimator of mes G obtained from some independent sample and such that

$$\varepsilon_{1,n}^2 = \sup_{G\in\mathcal{G}_{0,N}} E_G\left((\hat{\mu}_{1,n} - \text{mes } G)^2\right) \to 0$$

as $n\to\infty$. Then the estimator

$$\hat{\mu}_{2,n} = \int_K I\{x_N \leq \hat{g}^{odd}(x_1,\ldots,x_{N-1}) - C_*\psi_n\}dx +$$

$$+ \frac{2}{n}\hat{\mu}_{1,n} \sum_{i \text{ even}} I\{X_i\in\tilde{U}_0\}$$

has the maximal error

$$\varepsilon_{2,n}^2 = \sup_{G\in\mathcal{G}_{0,N}} E_G\left((\hat{\mu}_{2,n} - \text{mes } G)^2\right) = O\left(\max\{\psi_n/n; \varepsilon_{1,n}^2\psi_n^2\}\right)$$

as $n\to\infty$ (cf. (8.29)). The estimator $\hat{\mu}_{2,n}$ is either more accurate than the initial one $\hat{\mu}_{1,n}$ or $\varepsilon_{2,n}=O(\psi_n/n)$. Using $\hat{\mu}_{2,n}$ instead of $\hat{\mu}_{1,n}$, we get a new estimator $\hat{\mu}_{3,n}$, etc. After a finite number of steps the current estimator has such a small error that, if it is substituted into (8.28) instead of mes G_n^{odd}, it does not give rise to any additional error term with respect to $O(\psi_n/n)$.

To realize this idea, divide the whole sample $(X_1,\ldots,X_n)$

into k_0, $k_0>1$, independent subsamples $\mathcal{X}_k$, $k=1,\dots,k_0$, of size n/k_0; n/k_0 is assumed to be an even integer. Suppose that for k given, $1<k\le k_0$, there exists an estimator $\hat{\mu}_{k-1,n}$ of mes G measurable with respect to $\mathcal{X}_1,\dots,\mathcal{X}_{k-1}$, and such that

$$(8.36)\qquad \varepsilon^2_{k-1,n} = \sup_{G\in\mathcal{G}_{0,N}} E_G\Big((\hat{\mu}_{k-1,n} - \text{mes } G)^2\Big) \to 0$$

as $n\to\infty$. To define a pilot estimator $\hat{S}_{k,n}$ of the functional (8.25) we use $\hat{\mu}_{k-1,n}$ and the subsample $\mathcal{X}_k$ only. Again, split $\mathcal{X}_k$ into two subsamples containing odd and even observations, and let $\hat{g}^{odd}_k(x_1,\dots,x_{N-1})$ be the optimal edge estimator based on the subsample of $\mathcal{X}_k$ with odd indices. The estimator $\hat{S}_{k,n}$ is defined now by

$$(8.37)\qquad \hat{S}_{k,n} = \int_K \varphi(x) I\{x_N \le \hat{g}^{odd}_k(x_1,\dots,x_{N-1}) - C_*\psi_n\}dx +$$

$$+ 2(k_0/n)\,\hat{\mu}_{k-1,n} \sum_{\substack{i \text{ even} \\ X_i\in\mathcal{X}_k}} I\{X_i\in U_0\}\varphi(X_i)$$

where

$$U_0 = \Big\{x=(x_1,\dots,x_N)\in K:\ |x_N-\hat{g}^{odd}_k(x_1,\dots,x_{N-1})| \le C_*\psi_n\Big\}.$$

The following lemma is a direct extension of Theorem 8.6.1.

8.6.2. LEMMA. Let assumptions of Theorem 8.6.1 hold. Then

$$\sup_{G\in\mathcal{G}_{0,N}} E_G\Big((\hat{S}_{k,n}-S(G))^2\Big) = O\Big(\max\{\psi_n/n;\varepsilon^2_{k-1,n}\psi_n^2\}\Big),\quad n\to\infty.$$

Now we deduce a useful consequence.

8.6.3. LEMMA. Let assumptions of Theorem 8.6.1 hold and let $k_0=4+(N-1)/\gamma$. There exists an estimator μ^*_n of mes G measurable with respect to $\mathcal{X}_1,\dots,\mathcal{X}_{k_0-1}$ such that

$$\sup_{G\in\mathcal{G}_{0,N}} E_G\Big((\mu^*_n - \text{mes } G)^2\Big) = O(\psi_n/n),\quad n\to\infty.$$

Proof. Let $\hat{\mu}_{1,n}$ be the estimator of mes G obtained from the first subsample $\mathcal{X}_1$ and such that $\varepsilon^2_{1,n}=\psi^2_n$. Such an estimator exists due to Theorem 8.6.1. Put in Lemma 8.6.2 k=2 and $\varphi(x)\equiv 1$, so that S(G) = mes G. Then, in view of Lemma 8.6.2, the estimator $\hat{S}_{2,n}$ has the maximal error $O(\max\{\psi_n/n;\psi_n^4\})$, $n\to\infty$. Define $\hat{\mu}_{2,n}=\hat{S}_{2,n}$. If $\psi_n/n\geq\psi_n^4$ then put $\mu^*_n=\hat{\mu}_{2,n}$, and the lemma follows. Otherwise, repeat this procedure using the subsample $\mathcal{X}_3$ and taking into account that $\varepsilon^2_{2,n}=O(\psi_n^4)$, etc. After k_0-1 steps we have

$$\varepsilon^2_{k_0-1,n} = O\left(\max\{\psi_n/n;\psi_n^{2k_0-2}\}\right) = O(\psi_n/n), \quad n\to\infty,$$

due to the choice of k_0, and the lemma follows for the estimator $\mu^*_n=\hat{\mu}_{k_0-1,n}$. □

Let the integer k_0 and the estimator μ^*_n be defined by Lemma 8.6.3. Take $S^*_n=S_{k_0,n}$.

8.6.4. THEOREM. The estimator S^*_n of the functional (8.25) with a bounded Borel function $\varphi(x)$ satisfies

$$(8.38) \qquad \sup_{G\in\mathcal{G}_{0,N}} E_G\left((S^*_n-S(G))^2\right) = O(n^{-2\rho}(\log n)^{\gamma/(\gamma+N-1)}), \quad n\to\infty,$$

where $\rho=(\gamma+(N-1)/2)/(\gamma+N-1)$.

Proof. By Lemmas 8.6.2 and 8.6.3 the left-hand side of (8.38) has the order $\max\{\psi_n/n;(\psi_n/n)^2\psi_n^2\} = \psi_n/n$, $n\to\infty$. □

8.6.5. REMARK. Let us make some remarks concerning Theorem 8.6.4. First, note that it gives the upper bound on the minimax risk for estimators of S(G). The related lower bound is also valid which is exactly the same as in Theorems 8.3.1, 8.5.2 , and it is proved by similar means. In particular, the minimax rate in estimation of S(G) on convex domains with twice ($\gamma=2$) continuously differentiable edges equals $n^{-\rho}$, $\rho=(N+3)/(2(N+1))$ (up to the log-factor). For N=2 the exponent value equals $\rho=5/6$. In particular, this entails that the rate $n^{-2/3}$ of Rényi and Sulanke (1964) is not optimal (cf. Remark 7.5.1).

CHAPTER 9. IMAGE ESTIMATION FROM INDIRECT OBSERVATIONS

Let f(x) be an unknown smooth image function defined on a domain $G \subset R^N$. Assume that some transform $\mathcal{R}f(z)$ of the image f is available at given points $z_1,\dots,z_n$, i.e. that the observations are of the form

$$Y_i = \mathcal{R}f(z_i) + \xi_i, \qquad i = 1,\dots,n. \tag{9.1}$$

The random errors ξ_i are supposed to be i.i.d. $(0,\sigma^2)$-Gaussian random variables. We do not specify now the nature of the transform $\mathcal{R}f$. In the examples below $\mathcal{R}f(z)$ is a real-valued function defined on a domain $\mathcal{Z}$ in the space R^N of the same dimension as G.

How should the image function f(x) be estimated from the <u>indirect observations</u> (9.1)? First of all, we estimate the transform $\mathcal{R}f(z)$ by an estimator $\widehat{\mathcal{R}f} = \widehat{\mathcal{R}f}(z)$, then we try to invert it in some sense, i.e. to find the function $\hat{f} = \mathcal{R}^{-1}(\widehat{\mathcal{R}f})$. As a rule, the inversion can be done only approximately, by means of regularization techniques. This deteriorates the rate of convergence as compared to the case of direct (non-transformed) observations, and the minimax rates of convergence appear to be quite different.

Although the problem statement differs slightly from that discussed earlier, we consider it in this book since both approach and results show high resemblance with those developed in previous chapters.

We deal here with two specific problems: the blurred image estimation (Sections 9.1-9.3), and the stochastic problem of tomography (Sections 9.4-9.5). In the first one we follow Hall and Koch (1990). For technical reasons we give only a brief sketch, applying their approach to the approximation by Fourier series in L_2-norm. The problem of tomography has been studied in Korostelev and Tsybakov (1991) and its presentation here is also concise.

9.1. THE BLURRED IMAGE MODEL: ONE-DIMENSIONAL CASE

Let the image function $f: R^1 \to R^1$ be a smooth periodic function with period 2π. Denote by $\mathbf{f}_k$ the Fourier coefficients of f, i.e.

$$f(x) = \sum_{k=-\infty}^{\infty} \mathbf{f}_k \, e^{ikx} ,$$

$$\mathbf{f}_k = (2\pi)^{-1} \int_{-\pi}^{\pi} f(x) e^{-ikx} dx.$$

9.1.1. ASSUMPTION. There exist such $\beta \ge 1$ and $L>0$ that for each image function f its Fourier coefficients satisfy

$$\sum_{k=-\infty}^{\infty} |\mathbf{f}_k|^2 \, (1 + |k|^{2\beta}) \le L .$$

The class of images for which this assumption holds is denoted by $\mathbf{H}(\beta)=\mathbf{H}(\beta,L)$. We call B(x) a <u>function of systematic degradation</u> of the true image f and define the <u>blurred image</u> by

$$\mathcal{R}f(z) = (2\pi)^{-1} \int_{-\pi}^{\pi} f(x) B(z-x) dx, \quad z \in R^1. \tag{9.2}$$

9.1.2. ASSUMPTION. The function $B(x)$, $x \in R^1$, is a periodic function having period 2π and its Fourier coefficients $\mathbf{b}_k$ satisfy

$$c_1 |k|^{-æ} \le |\mathbf{b}_k| \le c_2 |k|^{-æ} , \quad k=0,\pm 1,\dots,$$

for some positive æ, c_1, c_2.

9.1.3. EXAMPLE. Let $B(x)=\pi^2-x^2$ for $-\pi<x<\pi$, then

$$\mathbf{b}_0=2\pi^2/3, \ \mathbf{b}_k=(-1)^{k-1}(2/k^2), \ k\neq 0.$$

In this example Assumption 9.1.2. holds with æ=2.

Suppose now that the indirect observations in (9.1) are generated by the values of the blurred image (9.2) measured at points z_i with the additive random noises. Let the design points $z_i, i=1,\dots,n$, be independent and uniformly distributed on $[-\pi,\pi]$. Introduce the L_2-norm

$$\|f(\cdot)\|_2 = \left((2\pi)^{-1}\int_{-\pi}^{\pi} |f(x)|^2 dx \right)^{1/2}$$

and consider the risk function

$$\mathrm{MISE}(\hat{f}_n, f) = E_f(\|\hat{f}_n(\cdot)-f(\cdot)\|_2^2)$$

where E_f is the expectation with respect to the joint distribution of (Y_i, z_i), $i=1,\dots,n$, satisfying (9.1), and $\hat{f}_n$ is an arbitrary estimator. Following the course of this book we want to find the minimax rate of convergence ψ_n which guarantees a non-degenerate asymptotics of the maximal risk value

$$\sup_{f\in\mathbf{H}(\beta)} \psi_n^{-2}\mathrm{MISE}(\hat{f}_n, f).$$

Denote by $\mathbf{r}_k$ the Fourier coefficients of the blurred image $\mathcal{R}f$:

$$\mathcal{R}f(z) = \sum_{k=-\infty}^{\infty} \mathbf{r}_k\, e^{ikz}, \qquad \mathbf{r}_k = (2\pi)^{-1}\int_{-\pi}^{\pi} \mathcal{R}f(z)e^{-ikz}dz. \tag{9.3}$$

The following lemma presents a well-known property of the convolution.

9.1.4. LEMMA. The Fourier coefficients

$$\mathbf{r}_k = \mathbf{f}_k\mathbf{b}_k, \qquad k=0,\pm 1,\dots . \tag{9.4}$$

<u>Proof</u>. It follows from (9.2) that

$$\mathcal{R}f(z) = (2\pi)^{-1}\int_{-\pi}^{\pi}\left(\sum_{k=-\infty}^{\infty} \mathbf{f}_k\, e^{ikx}\right)\left(\sum_{\ell=-\infty}^{\infty} \mathbf{b}_\ell\, e^{i\ell(z-x)}\right) dx =$$

$$= (2\pi)^{-1} \sum_{k,\ell=-\infty}^{\infty} \mathbf{f}_k \mathbf{b}_\ell e^{ikz} \int_{-\pi}^{\pi} (e^{ikx} - e^{i\ell x})\, dx = \sum_{k=-\infty}^{\infty} \mathbf{f}_k \mathbf{b}_k e^{ikz}.$$

Comparing this with (9.3), we get the lemma. □

We assume the function B is known to the statistician as well as its Fourier coefficients $\mathbf{b}_k$. The true image f can be represented by the series

$$f(x) = \sum_{k=-\infty}^{\infty} (\mathbf{r}_k / \mathbf{b}_k)\, e^{ikx} . \tag{9.5}$$

(surely, (9.5) has sense only if $\mathbf{b}_k \neq 0$). This suggests <u>the orthogonal series estimator</u> for f(x) which is defined as

$$f_n^*(x) = \sum_{|k| \le k_n} (\hat{\mathbf{r}}_k / \mathbf{b}_k)\, e^{ikx} \tag{9.6}$$

with some integer k_n, $k_n \to \infty$ as $n \to \infty$. Here

$$\hat{\mathbf{r}}_k = n^{-1} \sum_{\ell=1}^{n} Y_\ell \exp(-ikz_\ell) . \tag{9.7}$$

<u>Bias and variance</u>. The problem of the optimal choice of k_n can be solved, as usual, balancing the deterministic bias corresponding to $f_n^*(x)$ and its variance (cf. Section 1.6).

9.1.5. LEMMA.

(i) The bias $b(x) = E_f(f_n^*(x)) - f(x)$ satisfies

$$\|b(\cdot)\|_2^2 = (2\pi)^{-1} \int_{-\pi}^{\pi} |b(x)|^2 dx = \sum_{|k| > k_n} |\mathbf{f}_k|^2 ; \tag{9.8}$$

(ii) The integrated variance of the estimator (9.6) satisfies inequality

$$\mathrm{Var} = E_f(\|f_n^*(\cdot) - E_f(f_n^*(\cdot))\|_2^2) \le \tag{9.9}$$

$$\le \left(\sigma^2 + \|\mathcal{R}f(\cdot)\|_2^2\right) n^{-1} \sum_{|k|\le k_n} |\mathbf{b}_k|^{-2}.$$

<u>Proof.</u> Substituting (9.7) into (9.6) and taking into account (9.1), we have

$$b(x) = E_f\left(\sum_{|k|\le k_n} \left[(\mathbf{b}_k n)^{-1} \sum_{\ell=1}^{n} \left(\mathcal{R}f(z_\ell) + \xi_\ell\right) \exp(-ikz_\ell)\right] e^{ikx}\right) - \sum_{k=-\infty}^{\infty} \mathbf{f}_k e^{ikx} =$$

$$= \sum_{|k|\le k_n} \mathbf{b}_k^{-1} \left((2\pi)^{-1} \int_{-\pi}^{\pi} \mathcal{R}f(z)\exp(-ikz)\,dz\right) e^{ikx} - \sum_{k=-\infty}^{\infty} \mathbf{f}_k e^{ikx} =$$

$$= \sum_{|k|\le k_n} (\mathbf{r}_k/\mathbf{b}_k)\, e^{ikx} - \sum_{k=-\infty}^{\infty} \mathbf{f}_k e^{ikx} = - \sum_{|k|>k_n} \mathbf{f}_k\, e^{ikx}.$$

Hence (9.8) follows. To prove (9.9) note that

$$f_n^*(x) - E_f(f_n^*(x)) = \tilde{\Delta}_1(x) + \tilde{\Delta}_2(x)$$

where

$$\tilde{\Delta}_1(x) = \sum_{|k|\le k_n} \left[(\mathbf{b}_k n)^{-1} \sum_{\ell=1}^{n} \xi_\ell \exp(-ikz_\ell)\right] e^{ikx}$$

and

$$\tilde{\Delta}_2(x) = \sum_{|k|\le k_n} \left[(\mathbf{b}_k n)^{-1} \sum_{\ell=1}^{n} \left(\mathcal{R}f(z_\ell)\exp(-ikz_\ell\right) - \mathbf{f}_k\right] e^{ikx}.$$

Due to the independence of ξ_ℓ and z_ℓ the contribution of each term into the value of variance can be evaluated separately. The first term gives

$$E_f(|\tilde{\Delta}_1(x)|^2) = E_f\left(\sum_{|k|,|k'|\le k_n} \left[(\mathbf{b}_k\bar{\mathbf{b}}_{k'}n^2)^{-1} \sum_{\ell,\ell'=1}^{n} \xi_\ell \xi_{\ell'} \exp(-ikz_\ell + ik'z_{\ell'}) \right] \right) e^{ikx} e^{-ik'x} =$$

$$= \sum_{|k|,|k'|\le k_n} \frac{\sigma^2}{\mathbf{b}_k\bar{\mathbf{b}}_{k'}n} \left((2\pi)^{-1} \int_{-\pi}^{\pi} e^{-ikz+ik'z} dz \right) e^{ikx} e^{-ik'x} =$$

$$= \sum_{|k|\le k_n} \frac{\sigma^2}{|\mathbf{b}_k|^2 n}.$$

It means that $E_f(|\tilde{\Delta}_1(x)|^2)$ does not depend on x and

$$(9.10) \qquad E_f\left((2\pi)^{-1} \int_{-\pi}^{\pi} |\tilde{\Delta}_1(x)|^2 dx \right) = (\sigma^2/n) \sum_{|k|\le k_n} |\mathbf{b}_k|^{-2} .$$

Further,

$$E_f(|\tilde{\Delta}_2(x)|^2) = E_f\left(\sum_{|k|,|k'|\le k_n} \left[(\mathbf{b}_k\bar{\mathbf{b}}_{k'}n^2)^{-1} \sum_{\ell,\ell'=1}^{n} \mathcal{R}f(z_\ell)\overline{\mathcal{R}f}(z_{\ell'}) \exp(-ikz_\ell + ik'z_{\ell'}) - \mathbf{f}_k\bar{\mathbf{f}}_{k'} \right] \right) e^{ikx} e^{-ik'x} =$$

$$= \sum_{|k|,|k'|\le k_n} \left[(\mathbf{b}_k\bar{\mathbf{b}}_{k'}n^2)^{-1} (n^2-n)\mathbf{r}_k\bar{\mathbf{r}}_{k'} + \right.$$

$$+ (\mathbf{b}_k\bar{\mathbf{b}}_{k'},n)^{-1}(2\pi)^{-1}\int_{-\pi}^{\pi}|\mathcal{R}f(z)|^2\exp(-ikz+ik'z)dz - \mathbf{f}_k\bar{\mathbf{f}}_{k'}\Big] e^{ikx}e^{-ik'x}.$$

If we integrate the last expression in x the only terms corresponding to k = k′ are non-zero:

$$(9.11)\qquad E_f\left((2\pi)^{-1}\int_{-\pi}^{\pi}|\tilde{\Delta}_2(x)|^2dx\right) =$$

$$= \sum_{|k|\le k_n}\left[(1-1/n)|\mathbf{f}_k|^2 + (\|\mathcal{R}f(\cdot)\|_2^2/n)|\mathbf{b}_k|^{-2}-|\mathbf{f}_k|^2\right] \le$$

$$\le (\|\mathcal{R}f(\cdot)\|_2^2/n)\sum_{|k|\le k_n}|\mathbf{b}_k|^{-2}.$$

Relations (9.10) and (9.11) entail (9.9). This proves the lemma.□

9.2. HIGH-DIMENSIONAL BLURRED IMAGE MODELS

The N-dimensional extension of the previous section model appears if we assume that the image function $f:R^N\to R^1$, is a smooth 2π-periodic function in all the arguments, $N\ge 1$. Let $k=(k_1,\dots k_N)$ be a multi-index of integers, $|k|=|k_1|+\dots+|k_N|$, and let by definition

$$<k,x>=k_1x_1+\dots+k_Nx_N, \text{ for } x\in R^N.$$

Denote by $\mathbf{f}_k$ the Fourier coefficients of f, i.e.

$$f(x) = \sum_{|k|=0}^{\infty}\mathbf{f}_k\exp(i<k,x>)\ ,\ x\in R^N,$$

$$\mathbf{f}_k = (2\pi)^{-N}\int_{[-\pi,\pi]^N} f(x)\exp(-i<k,x>)dx.$$

On the analogy of 9.1.1 the class of images $\mathbf{H}_N(\beta)=\mathbf{H}_N(\beta,L)$,

$\beta>N/2$, is defined by

$$\mathbf{H}_N(\beta,L) = \left\{ f(x),\ x\in R^N:\ \sum_{|k|=0}^{\infty} |\mathbf{f}_k|^2(1+|k|^{2\beta}) \le L \right\}.$$

The function $B(x)$, $x\in R^N$, is a <u>function of systematic degradation</u> if it is 2π-periodic in all the arguments and its Fourier coefficients satisfy

9.2.1. ASSUMPTION. The Fourier coefficients $\mathbf{b}_k$ of $B(x)$, $x\in R^N$, are such that

$$c_1|k|^{-æ} \le |\mathbf{b}_k| \le c_2|k|^{-æ}, \quad |k|=0,1,\ldots,$$

for some positive æ, c_1, c_2.

The N-dimensional blurred image model is defined by

$$\mathcal{R}f(z) = (2\pi)^{-N} \int_{[-\pi,\pi]^N} f(x)B(z-x)dx,\ z\in R^N$$

(cf. (9.2)).

9.2.2. ASSUMPTION. The design points z_i in observations (9.1) are independent uniformly distributed on $[-\pi,\pi]^N$, and the design $\{z_i,\ i=1,\ldots,n\}$ is independent of errors $(\xi_1,\ldots,\xi_n)$.

Introduce the L_2-norm

$$\|f(\cdot)\|_2 = \left((2\pi)^{-N} \int_{[-\pi,\pi]^N} |f(x)|^2dx \right)^{1/2}$$

and consider <u>the orthogonal series estimator</u>:

$$f_n^*(x) = \sum_{|k|\le k_n} (\hat{\mathbf{r}}_k/\mathbf{b}_k)\ \exp(i\langle k,x\rangle),\ x\in R^N, \tag{9.12}$$

with some integer k_n, $k_n\to\infty$ as $n\to\infty$. Here

$$\hat{\mathbf{r}}_k = n^{-1} \sum_{\ell=1}^{n} Y_\ell \exp(-i\langle k z_\ell \rangle).$$

The risk function

$$\text{MISE}(\hat{f}_n, f) = E_f(\|\hat{f}_n(\cdot) - f(\cdot)\|_2^2)$$

is the same as above for N-dimensional models.

Minimax rates of convergence. Since no specific features of the one-dimensional model have been used in the proof of Lemma 9.1.5, its statement remains valid for high-dimensional blurred images. We apply it to obtained the rate of convergence.

9.2.3. THEOREM. Let Assumptions 9.2.1 and 9.2.2 hold. Choose $k_n = n^{1/(2\beta+2æ+N)}$ in (9.12) . Then

$$\sup_{f \in \mathbf{H}_N(\beta, L)} \text{MISE}(f_n^*, f) \le C_1 \, n^{-2\beta/(2\beta+2æ+N)}$$

with some positive C_1.

Proof. It follows from Lemma 9.1.5 that

$$\text{MISE}(f_n^*, f) = \|b(\cdot)\|_2^2 + \text{Var} \le \tag{9.13}$$

$$\le c_3 \Big(\sum_{|k|>k_n} |\mathbf{f}_k|^2 + n^{-1} \sum_{|k| \le k_n} |\mathbf{b}_k|^{-2} \Big)$$

for some positive constant c_3. The definition of the class $\mathbf{H}_N(\beta, L)$ guarantees that for some positive c_4

$$\sum_{|k|>k_n} |\mathbf{f}_k|^2 \le \Big(\sum_{|k|>k_n} |\mathbf{f}_k|^2 (1 + |k|^{2\beta}) \Big) / (1 + k_n^{2\beta}) \le c_4 k_n^{-2\beta}. \tag{9.14}$$

Assumption 9.2.1 implies

$$\sum_{|k|\le k_n} |\mathbf{b}_k|^{-2} \le \sum_{|k|\le k_n} c_1^{-1}|k|^{2æ} \le c_5\, k_n^{2æ+N}. \tag{9.15}$$

for some positive c_5. Substituting (9.14), (9.15) into (9.13), we finally have

$$\mathrm{MISE}(f_n^*, f) \le c_3\left(c_4\, k_n^{-2\beta} + c_5\, n^{-1}\, k_n^{2æ+N} \right) \le C_1\, n^{-2\beta/(2\beta+2æ+N)}$$

where $C_1 = c_3(c_4+c_5)$. □

9.2.4. THEOREM. If Assumptions 9.2.1 and 9.2.2 hold, then for the problem of blurred image estimation from indirect observations (9.1) the following lower bound is true

$$\liminf_{n\to\infty}\ \inf_{T_n}\ \sup_{f\in\mathbf{H}_N(\beta,L)} \left(n^{2\beta/(2\beta+2æ+N)}\mathrm{MISE}(T_n, f)\right) \ge C_0$$

where $C_0>0$ is a constant. Here $\inf_{T_n}$ denotes the infimum over all estimators.

Proof of this theorem is based on standard application of Assouad's lemma (Theorem 2.6.4), with the functions $\vartheta_j = f_{jn}(\cdot)$ defined in (2.53) ($\delta_n = [n^{1/(2\beta+2æ+N)}]^{-1}$).

9.2.5. REMARK. The case where $B(x)=\delta_0(x)$ is the delta-function (and thus the image f in (9.2) is not blurred) corresponds formally to $æ=0$ in Assumption 9.1.2. Hence, Theorems 9.2.3 and 9.2.4 are in the good accordance with the results on estimation of the regression function in L_2-norm (cf. Theorem 1.6.8 and Example 2.7.5). Note that the rate of convergence is worse for blurred images since

$$2\beta/(2\beta+2æ+N) < 2\beta/(2\beta+N)$$

if $æ>0$ (cf. Section 2.7.6).

9.3. UPPER BOUNDS IN NON-REGULAR CASE

Unfortunately, Assumption 9.2.1 is too restrictive as well as its one-dimensional version 9.1.2. For example, the Fourier coefficients of the function $B(x)=I(|x|\le\tau)$, $0<\tau<\pi$, are equal to

$\mathbf{b}_k=\pi^{-1}\sin k\tau$, $k=\pm1,\pm2,\ldots$; here N=1 and $x\in R^1$. Hence, they are not separated from zero. Moreover, they are exactly zero for some k if τ/π is rational. It means that (9.5) and (9.6) cannot be applied.

We define a <u>non-regular case</u> as the one when the lower bound for $|\mathbf{b}_k|$ is violated in Assumption 9.2.1. In this case the estimator (9.12) can be modified:

$$(9.16) \qquad f_n^*(x) = \sum_{k\notin\Psi_n} (\hat{\mathbf{r}}_k/\mathbf{b}_k)\exp(i\langle k,x\rangle), \quad x\in R^N,$$

where Ψ_n is some set of integers. How should this set be chosen? Similarly to (9.13) the integrated squared risk

$$(9.17) \qquad \text{MISE}(f_n^*,f) \le c_6\Big(\sum_{k\in\Psi_n} |\mathbf{f}_k|^2 + n^{-1}\sum_{k\notin\Psi_n} |\mathbf{b}_k|^{-2}\Big), \quad c_6>0.$$

Note that the estimator (9.16) can be inconsistent. Indeed, if $\mathbf{b}_k=0$ for some k then the Fourier coefficient $\mathbf{f}_k$ cannot be found from observations (9.1) even in the absence of noises ξ_i. Nevertheless, we would like to choose Ψ_n which guarantees some upper bound for MISE.

9.3.1. ASSUMPTION. Let the Fourier coefficients of the image function $f:R^N\to R^1$, satisfy

$$|\mathbf{f}_k| \le \gamma_k, \quad |k|=0,1,\ldots,$$

for a given positive sequence γ_k such that $\sum_{|k|\ge0} \gamma_k^2 < \infty$.

Assumption 9.3.1 permits us to give a simple rule to choose Ψ_n in (9.16) and the explicit upper bound for the integrated squared risk (9.17). Define for any positive value ε

$$\Psi(\varepsilon) = \Big\{ k:\ \gamma_k^2|\mathbf{b}_k|^2 \le \varepsilon \Big\},$$

and let $\Psi_n = \Psi(\varepsilon_n)$ where ε_n is defined as the smallest positive value satisfying the following inequality:

$$\sum_{k\in\Psi(\varepsilon_n)} \gamma_k^4 \, |\mathbf{b}_k|^2 \ge 1/n. \tag{9.18}$$

9.3.2. THEOREM. Let Assumption 9.3.1 hold and let Ψ_n be chosen in accordance with (9.18). Then there exists a positive C_1 such that

$$\mathrm{MISE}(f_n^*, f) \le C_1 \sum_{k\in\Psi_n} \gamma_k^2 \tag{9.19}$$

Proof. Assume without loss of generality that $\sum_{|k|\ge 0} \gamma_k^2 = 1$. Let W be a discrete random variable having the distribution

$$P(W = \gamma_k^2 |\mathbf{b}_k|^2) = \gamma_k^2, \quad |k|=0,1,\ldots$$

Note that it is sufficient to prove the inequality

$$\left(\sum_{k\in\Psi_n} \gamma_k^4 \, |\mathbf{b}_k|^2 \right) \left(\sum_{k\notin\Psi_n} |\mathbf{b}_k|^{-2} \right) \le c_7 \sum_{k\in\Psi_n} \gamma_k^2 \tag{9.20}$$

for a positive constant c_7. Indeed, (9.20) together with (9.17) and Assumption 9.3.1 entails (9.19). Let us write the left-hand side of (9.20) in terms of the random variable W :

$$E\Big(W \, I(W \le \varepsilon_n)\Big) \, E\Big(W^{-1} I(W > \varepsilon_n)\Big) \le \varepsilon_n P\Big(W \le \varepsilon_n\Big) \, E\Big(W^{-1} I(W > \varepsilon_n)\Big) \le$$

$$\le P\Big(W \le \varepsilon_n\Big) E\Big((\varepsilon_n/W) \, I(\varepsilon_n/W < 1)\Big) \le P\Big(W \le \varepsilon_n\Big).$$

This proves the lemma since $P\Big(W \le \varepsilon_n\Big) = \sum_{k\in\Psi_n} \gamma_k^2$. □

9.3.3. REMARK. The upper bound in (9.19) is not the minimax one as that in Theorem 9.2.3. Indeed, let N=1, and let Assumption 9.1.2 and the following inequalities hold

(9.21) $$|f_k| \le |k|^{-\beta-1/2}, \quad k=0,\pm1,\dots .$$

(Condition (9.21) is close to Assumption 9.1.1.)

Then $\Psi_n=\Psi(\varepsilon_n)$ should be chosen in accordance with the rule:

$$\sum_{k\in\Psi_n} |k|^{-(4\beta-2)-2æ} = n^{-1}(1+o(1)) \text{ as } n\to\infty, \text{ and } |k|^{-(2\beta-1)-2æ} \le \varepsilon_n ,$$

i. e. $\Psi_n = \left\{k: |k| \ge c_8 n^{1/(4\beta+2æ+1)}\right\}$, $c_8>0$. Though Theorems 9.2.3 and 9.2.4 do not apply directly to the class of images satisfying (9.21), it is easy to show that this choice is not the best, i.e. the corresponding estimator (9.16) is not the optimal one.

9.4. THE STOCHASTIC PROBLEM OF TOMOGRAPHY

The rest of this chapter is devoted to a stochastic problem of tomography which is another example of image estimation from indirect data. Let $f(x)$: $R^N \to R^1$, be an unknown image function. A priori information on the image function $f(x)$ is given by:

9.4.1. ASSUMPTION. The support of f is contained in the unit ball: $\text{supp } f \subseteq \{x: |x|\le1\}$, i.e. $f(x)=0$ if $|x|\ge1$.

9.4.2. ASSUMPTION. The Fourier transform of f,

$$\tilde{f}(\omega) = \int_{R^N} \exp(i(x,\omega))f(x)dx, \quad \omega\in R^N,$$

satisfies

$$\int_{R^N} (1 + |\omega|^2)^{\beta} |\tilde{f}(\omega)|^2 d\omega \le L$$

for some positive L and $\beta>N/2$.

Let β, L be fixed. Denote by $\tilde{\mathbf{H}}(\beta,L)$ the class of functions f satisfying Assumptions 9.4.1 and 9.4.2. Since $\beta>N/2$ the functions belonging to $\tilde{\mathbf{H}}(\beta,L)$ are continuous; this follows from the embedding theorems for Sobolev spaces (see Yosida (1965), sec. 6.7).

9.4.3. DEFINITION. Let $S^{N-1} = \{x: |x|=1\}$ be the unit sphere in R^N and let $(\mathbf{s},u)$, $\mathbf{s}\in S^{N-1}$, $|u|\le 1$, be a point in the cylinder

$$\mathcal{Z} = S^{N-1} \times [-1,1].$$

The Radon transform of f is the following function on $\mathcal{Z}$:

$$\mathcal{R}f(\mathbf{s},u) = \int_{x:(x,\mathbf{s})=u} f(x)dx.$$

Assume that $(\mathbf{s}_i,u_i)$, $i=1,\dots,n$, are some points in $\mathcal{Z}$ and let observations be of the form

$$Y_i = \mathcal{R}f(\mathbf{s}_i,u_i) + \xi_i, \quad i=1,\dots,n. \tag{9.22}$$

where ξ_i are i.i.d. random variables with zero mean and finite positive variance. To simplify the statement of results suppose that the distribution of noise is Gaussian keeping in mind the redundancy of this assumption (see Korostelev and Tsybakov (1991)).

9.4.4. ASSUMPTION. The design points $(\mathbf{s}_i,u_i)$, $i=1,\dots,n$, are independent uniformly distributed on $\mathcal{Z}$; these points are independent of errors $(\xi_1,\dots,\xi_n)$ in (9.22).

Let us study the asymptotics of minimax risk in image estimation from observations (9.22). The problem is to find minimax rates of convergence ψ_n on the classes of images $\tilde{\mathbf{H}}(\beta,L)$. The risk function of an estimator $\hat{f}_n(x)$ is

$$\mathrm{MISE}(\hat{f}_n,f) = E_f\left(\int (\hat{f}_n(x)-f(x))^2dx\right).$$

Here and later E_f is the expectation with respect to the joint distribution of the design points and observations $(\mathbf{s}_i,u_i,Y_i)$, $i=1,\dots,n$, satisfying (9.22).

If $\xi_i=0$, then (9.22) reduces to the well-known model that was extensively studied in the theory of inverse problems (see, e.g. Natterer (1980), Herman (1980), Natterer (1986)). The basic problem related to this deterministic model is to find (or to approximate) the inverse of the Radon transform. Some features of this problem are discussed below.

The problem of tomography as stated in this section is a "regression type" stochastic inverse problem. Also a "density type" stochastic problem of tomography is interesting for some applications as shown by Johnstone and Silverman (1990). They supposed that the image is a probability density function, and that the sample of i.i.d. observations $(\mathbf{s}_i, u_i)$ distributed with the density $\mathcal{R}f$ is given. Johnstone and Silverman (1990) proved that it is possible to estimate f from such observations and established the minimax rates of convergence in this density estimation type setting.

Deterministic problem. Suppose that the Radon transform $\mathcal{R}f(\mathbf{s},u)$ is a known function on $\mathcal{Z}$. How to restore the image function f? Here we discuss a well-known approach to this deterministic problem. Define a kernel

$$K_\delta(u) = 2\rho_N(2\pi)^{-N}\int_0^{1/\delta} r^{N-1}\cos(ur)\,dr,\quad u\in R^1,$$

depending on a positive parameter δ, where $\rho_N = \mathrm{mes}(S^{N-1}) = 2\pi^{N/2}/\Gamma(N/2)$ is the Lebesgue measure of the unit sphere S^{N-1}.

The Fourier transform of K_δ is

$$\tilde{K}_\delta(t) = (2\pi)^{1-N}\rho_N\,|t|^{N-1}I_\delta(t),\quad t\in R^1,$$

where $I_\delta(t)$ is the indicator-function:

$$I_\delta(t) = \begin{cases} 1 \text{ if } |t|<1/\delta, \\ 0 \text{ if } |t|\ge 1/\delta. \end{cases}$$

Let us show that the "kernel transform"

(9.23) $$f_\delta(x) = (2\rho_N)^{-1}\int_{\mathcal{Z}} K_\delta((\mathbf{s},x)-u)\mathcal{R}f(\mathbf{s},u)\,d\mathbf{s}du$$

reconstructs f as δ tends to 0. Use the following property of the Radon transform.

9.4.5. LEMMA. (The "projection theorem").

$$\hat{\mathcal{R}}f(\mathbf{s},t) = \tilde{f}(t\mathbf{s}),\ t\in R^1,\ \mathbf{s}\in S^{N-1},$$

where $\hat{\mathcal{R}}f(\mathbf{s},t)$ is the Fourier transform of $\mathcal{R}f(\mathbf{s},u)$ over the argument u only.

Proof. If $\mathbf{s}^\perp\in S^{N-1}$ and $(\mathbf{s},\mathbf{s}^\perp) = 0$, then

$$\hat{\mathcal{R}}f(\mathbf{s},t) = \int_{-\infty}^{\infty} \mathcal{R}f(\mathbf{s},u)e^{itu}du = \int_{-\infty}^{\infty}\left(\int_{-\infty}^{\infty} f(u\mathbf{s}+v\mathbf{s}^\perp)dv\right) e^{itu}du =$$

$$= \int_{-\infty}^{\infty}\int_{-\infty}^{\infty} f(u\mathbf{s}+v\mathbf{s}^\perp)\exp\{i(u\mathbf{s}+v\mathbf{s}^\perp,t\mathbf{s})\}dudv = \tilde{f}(t\mathbf{s}).\ \square$$

9.4.6. LEMMA. For every $x \in \{x:|x|\le 1\}$ the bias function $b_\delta(x)=f_\delta(x)-f(x)$, $f\in\tilde{\mathbf{H}}(\beta,L)$, can be presented in the form

$$b_\delta(x) = (2\pi)^{-N}\int_{R^N} (I_\delta(|\omega|) - 1)\tilde{f}(\omega)e^{-i(\omega,x)}d\omega\ .$$

Proof. Functions of the class $\tilde{\mathbf{H}}(\beta,L)$, $\beta>N/2$, are uniformly bounded and satisfy by definition the identity: $\mathcal{R}f(\mathbf{s},u)=0$ if $|u|>1$. Hence

$$f_\delta(x) = (2\rho_N)^{-1}\int_{S^{N-1}}\left[\int_{-1}^{1} K_\delta((\mathbf{s},x)-u)\mathcal{R}f(\mathbf{s},u)du\right]d\mathbf{s} =$$

$$= (2\rho_N)^{-1}\int_{S^{N-1}}\left[\int_{-\infty}^{\infty} K_\delta((\mathbf{s},x)-u)\mathcal{R}f(\mathbf{s},u)du\right]d\mathbf{s}\ .$$

Using the inverse Fourier transform and the properties of the convolution, one gets

$$\int_{\infty}^{\infty} K_\delta((\mathbf{s},x)-u)\mathcal{R}f(\mathbf{s},u)\,du = (2\pi)^{-1}\int_{-\infty}^{\infty} \tilde{K}_\delta(t)\widehat{\mathcal{R}f}(\mathbf{s},t)e^{-it(\mathbf{s},x)}dt =$$

$$= (2\pi)^{-N}\rho_N \int_{-\infty}^{\infty} |t|^{N-1} I_\delta(t)\tilde{f}(t\mathbf{s})e^{-i(t\mathbf{s},x)}dt$$

where the "projection theorem" (Lemma 9.4.5) has been applied. Thus,

$$f_\delta(x) = (1/2)\ (2\pi)^{-N} \int_{S^{N-1}} \left[\int_{-\infty}^{\infty} |t|^{N-1} I_\delta(t)\tilde{f}(t\mathbf{s})e^{-i(t\mathbf{s},x)}dt\right]d\mathbf{s} =$$

$$= (2\pi)^{-N} \int_{S^{N-1}} \int_{0}^{\infty} |t|^{N-1} I_\delta(t)\tilde{f}(t\mathbf{s})e^{-i(t\mathbf{s},x)}dt\,d\mathbf{s} =$$

$$= (2\pi)^{-N} \int_{R^N} I_\delta(|\omega|)\tilde{f}(\omega)\ e^{-i(\omega,x)}\ d\omega .$$

This proves the lemma.□

9.4.7. COROLLARY. For every x, $|x|\le 1$, and $f\in\tilde{\mathbf{H}}(\beta,L)$, we have $f_\delta(x)\to f(x)$ as $\delta\to 0$. The proof is straightforward since

$$(I_\delta(|\omega|)\ -\ 1)\tilde{f}(\omega)\ \to\ 0\ \text{as}\ \delta\ \to\ 0.$$

9.5. MINIMAX RATES OF CONVERGENCE

Optimal kernel estimators. In the stochastic problem of tomography under Assumption 9.4.4 it is natural to define a statistical estimator as the following sample analogue to the kernel transform (9.23):

$$(9.24)\qquad f_n^*(x) = n^{-1}\sum_{i=1}^{n} K_{\delta_n}((\mathbf{s}_i,x)\ -\ u_i)Y_i$$

where $\delta_n \to 0$ as $n \to \infty$.

The following lemma describes the bias and variance terms of the estimator (9.24).

9.5.1. LEMMA. Let Assumption 9.4.4 hold. Then for every x, $|x|\le 1$, the bias $b_n(x) = E_f\left(f_n^*(x)\right) - f(x)$ is represented by

$$b_n(x) = (2\pi)^{-N} \int_{R^N} (I_\delta(|\omega|)-1)\tilde{f}(\omega)e^{-i(\omega,x)}d\omega , \tag{9.25}$$

and variance $\sigma_n^2(x) = E_f\left((f_n^*(x) - E_f\left(f_n^*(x)\right))^2\right)$ satisfies

$$\sigma_n^2(x) \le c_9\, \delta_n^{-(2N-1)} n^{-1} \tag{9.26}$$

with some positive constant c_9 uniformly over $f\in\tilde{\mathbf{H}}(\beta,L)$.

Proof. The identity $E_f\left(f_n^*(x)\right) = f_{\delta_n}(x)$ holds under Assumption 9.4.4, and (9.25) follows from Lemma 9.4.6. Further, this assumption entails the equality

$$\sigma_n^2(x) = \sigma^2(2n\rho_N)^{-1} \int_{S^{N-1}} \int_{-1}^{1} K_{\delta_n}^2((\mathbf{s},x)-u)\,du\,d\mathbf{s} \tag{9.27}$$

and

$$\int_{-1}^{1} K_{\delta_n}^2((\mathbf{s},x)-u)du \le \int_{-\infty}^{\infty} K_{\delta_n}^2(u)du = (2\pi)^{-1}\int_{-\infty}^{\infty}|\tilde{K}_{\delta_n}(t)|^2dt =$$

$$= (2\pi)^{1-N}\rho_N^2 \int_{-\infty}^{\infty} |t|^{2N-2} I_{\delta_n}(t)dt \le c_{10}\, \delta_n^{-(2N-1)} .$$

Substituting the last quantity into (9.27), we prove (9.26).□

9.5.2. THEOREM. Let Assumption 9.4.4 be satisfied, $\beta>N/2$, and let $\delta_n=n^{-1/(2\beta+2N-1)}$ in (9.24). Then the upper bound holds

$$\sup_{f\in\tilde{\mathbf{H}}(\beta,L)} \psi_n^{-2}\, \mathrm{MISE}(f_n^*,f) \le C_1$$

with $\psi_n = n^{-\beta/(2\beta+2N-1)}$ and a positive C_1.

Proof. By (9.25) and the Parseval identity the squared L_2-norm of the bias is evaluated as follows:

$$\|b_n(\cdot)\|^2 = (2\pi)^{-2N} \int_{R^N} (I_\delta(|\omega|) - 1)^2 \, |\tilde{f}(\omega)|^2 \, d\omega \le$$

$$\le (2\pi)^{-2N} \int_{R^N} |\tilde{f}(\omega)|^2 (\delta|\omega|)^{2\beta} d\omega \le L(2\pi)^{-2N} \delta^{2\beta} .$$

Using this inequality and (9.26), one finds that the balance between the bias and variance terms is attained if δ_n is chosen as in the theorem. The rest of proof is standard (cf. Theorem 1.6.8). □

Image estimation at a fixed point. The optimality of the estimator (9.24) can be also proved with the risk function

$$\text{MSE}(\hat{f}_n, f, x_0) = E_f\left((\hat{f}_n(x_0) - f(x_0))^2\right) \tag{9.28}$$

where x_0 is a fixed point in the open ball $\{x: |x|<1\}$.

In view of Lemma 9.5.1 the bias of estimator (9.24) at point x_0 is evaluated as follows:

$$|b(x_0)| \le (2\pi)^{-N} \int_{R^N} |\tilde{f}(\omega)|^2 \left[2(\delta_n|\omega|)^{\beta}/(1+(\delta_n|\omega|)^{\beta})\right] d\omega \le$$

$$\le 2(2\pi)^{-N} \delta_n^{\beta} \left[\int_{R^N} |\tilde{f}(\omega)|^2 |\omega|^{2\beta} d\omega\right]^{1/2} \left[\int_{R^N} (1+(\delta_n|\omega|)^{\beta})^{-2} d\omega\right]^{1/2} \le$$

$$\le c_{11} \delta_n^{\beta-N/2}, \quad c_{11}>0.$$

The variance of $\hat{f}_n$ under the uniform random design is of the order $n^{-1}\delta_n^{-(2N-1)}$. Hence,

$$\text{MSE}(\hat{f}_n, f, x_0) = b^2(x_0) + \text{variance} \le c_{12}(\delta^{2\beta-N}) + c_{13}(n^{-1}\delta_n^{-(2N-1)}).$$

For $\delta_n = n^{-1/(2\beta+N-1)}$ the balance between the bias and variance terms is attained, and $\mathrm{MSE}(\hat{f}_n, f, x_0) = O(n^{-(2\beta-N)/(2\beta+N-1)})$. The rate obtained is the minimax rate of convergence for this specific risk. We omit details (see Korostelev and Tsybakov (1991)).

Lower bounds. The sequence ψ_n in Theorem 9.5.2 proves to be the minimax rate of convergence in the L_2-norm. Moreover, this rate cannot be improved by any design, not necessary by the uniformly distributed one.

9.5.3. THEOREM. Let $\{(\mathbf{s}_i, u_i),\ i=1,\ldots,n\}$ be an arbitrary design on $\mathcal{Z}$ (deterministic or random) independent of errors $(\xi_1,\ldots\xi_n)$ in (9.22). Then there exists a positive constant C_0 such that the inequality holds

$$\liminf_{n\to\infty} \inf_{\hat{f}_n} \sup_{f\in\tilde{\mathbf{H}}(\beta,L)} \psi_n^{-2}\mathrm{MISE}(\hat{f}_n, f) \geq C_0$$

for some constant $C_0>0$ and $\psi_n = n^{-\beta/(2\beta+2N-1)}$. Here $\inf_{\hat{f}_n}$ is the infimum over all estimators.

9.5.4. COROLLARY. Under the assumptions of Theorem 9.5.2 the S-minimax rate of convergence for $\tilde{\mathbf{H}}(\beta,L)$ in L_2-norm equals to $\psi_n = n^{-\beta/(2\beta+2N-1)}$ and the S-optimal estimator has the form (9.24).

Proof of Theorem 9.5.3 is given in Korostelev and Tsybakov (1991). It is based on Assouad's lemma which, however, cannot be applied here in a standard way (cf. Theorem 2.8.4). The obstacle is in the fact that for two test functions with non-overlapping supports the supports of the Radon transforms always overlap. This hurdle can be eliminated using some geometric argument.

REFERENCES

Abbar,H. (1990) Un estimateur spline du contour d'une répartition ponctuelle aléatoire. Statistique et Analyse des Données, 15, n.3, 1-19.

Assouad,P. (1983) Deux remarques sur l'estimation, C. R. Acad. Sci. Paris 296, 1021-1024.

Besov,O.V., Il'in,V.P., and Nikol'skii,S.M. (1975) Integral Representations of Functions and Embedding Theorems. Nauka, Moscow.

Birgé,L. (1983) Approximation dans les espaces métriques et la théorie de l'éstimation, Z. Wahrscheinlichkeitstheorie verw. Geb. 65, 181-238.

Birgé,L. and Massart,P. (1991) Rates of convergence for minimum contrast estimators, Rapport Technique # 140, L.S.T.A., Université Paris 6.

Blake,A. and Zisserman,A. (1987) Visual reconstruction. MIT Press, Cambridge, MA.

Bosq,D. (1971) Contribution à la théorie de l'estimation fonctionnelle. Thèse, Paris.

Bretagnolle,J. and Huber,C. (1979) Estimation des densités: risque minimax, Z. für Wahrscheinlichkeitstheorie und verw. Geb. 47, 119-137.

Carlstein,E. and Krishnamoorthy,C. (1992) Boundary estimation. J. of the American Statistical Association 87, n.418, 430-438.

Chevalier,J. (1976) Estimation du support et du contenu du support d'une loi de probabilité. Ann. Inst. H.Poincaré, sec. B., 12, n.4, 339-364.

Cleveland,W.S. (1979) Robust locally weighted regression and smoothing scatterplots. J. of the American Statistical Association 74, n.368, 829-836.

Deprins,D., Simar,L. and Tulkens,H. (1984) Measuring labor efficiency in post offices. In: "The performance of public enterprises: Concepts and Measurements", M.Marchand, P.Pestieau and H.Tulkens, eds. Amsterdam, North-Holland, 243-267.

Devroye,L. and Györfi,L. (1985) Nonparametric Density Estimation: the L_1-view, J.Wiley, N.Y.

Devroye,L. and Wise,G.L. (1980) Detection of abnormal behavior via nonparametric estimation of the support. SIAM J. Applied Math. 38, 480-488.

Dudley,R.M. (1974) Metric entropy of some classes of sets with differentiable boundaries. J.Approx.Theory 10, 227-236.

Farrell,M.J. (1957) The measurement of productive efficiency. J. Royal Statistical Soc., A, 120, 253-281.

Galambos,J. (1978) The Asymptotic Theory of Extreme Order Statistics. J.Wiley, N.Y. e.a.

Geffroy,J. (1964) Sur un problème d'estimation géométrique. Publications de l'Institut de Statistique des Universités de Paris, 13, 191-210.

Geman,S. and Geman,D. (1984) Stochastic relaxation, Gibbs distribution, and Bayesian restoration of images. IEEE Trans. on PAMI, 6, 721-741.

Girard,D. (1990) From template matching to optimal approximation by piecewise smooth curves. In: Curves and Surfaces in Computer Vision and Graphics, Proc. Conf. Santa Clara, Calif., 13-15 February, 1990, 174-182.

Grenander,U. (1976, 1978, 1981) Lectures in Pattern Theory, vol. 1-3. Springer, N.Y. e.a. (Applied Mathematical Sciences, vol. 18, 24, 33).

Hall,P. and Koch,I. (1990) On continuous image models and image analysis in the presence of correlated noise. Adv. Appl. Prob., 22, 332-349.

Haralick,R.M. (1980) Edge and region analysis for digital image data. Comput. Graphics and Image Processing 12, 60-73.

Hartigan,J.A. (1975) Clustering Algorithms. J.Wiley, N.Y. e.a.

Hartigan,J.A. (1987) Estimation of a convex density contour in two dimensions. J. of the American Statistical Association 82, n. 397, 267-270.

Härdle,W. (1990) Applied Nonparametric Regression, Cambridge Univ. Press.

Herman,G.T. (1980) Image Reconstruction from Projections, Academic Press, N.Y.

Hoeffding,W. (1963) Probability inequalities for sums of bounded random variables. J. of the American Statistical Association 58, n.301, 13-30.

Huang,J.S. and Tseng,D.H. (1988) Statistical theory of edge detection. Comput. Graphics and Image Processing 43, 337-346.

Ibragimov,I.A. and Khasminskii,R.Z. (1980a) On estimation of distribution density, Zapiski Nauchnych Seminarov LOMI, 98,61-86.

Ibragimov,I.A. and Khasminskii,R.Z. (1980b) On nonparametric estimation of regression, Doklady Acad. Nauk SSSR 252, 780-784.

Ibragimov,I.A. and Khasminskii,R.Z. (1981) Statistical Estimation: Asymptotic Theory, Springer-Verlag, N.Y..

Ibragimov,I.A. and Khasminskii,R.Z. (1982) On the bounds for quality of nonparametric regression function estimation, Theory of Probability and its Applications 27, 81-94.

Ibragimov,I.A., Nemirovskii,A.S. and Khasminskii,R.Z. (1986) Some problems of nonparametric estimation in Gaussian white noise, Theory of Probability and its Applications 31, 451-466.

Jacob,P. (1984) Estimation du contour discontinu d'un processus ponctuel sur le plan. Publications de l'Institut de Statistique des Universités de Paris, 29, 3-4, 1-25.

Jacob,P. and Abbar,H. (1989) Estimating the edge of Cox process area. Cahiers du Centre d'Etudes de Recherche Opérationnele, 31, n.3-4.

Johnson,N.L. and Kotz,S. (1970) Continuous univariate distributions - 2. Houghton Mifflin, Boston e.a.

Johnstone,I.M. and Silverman,B.W. (1990) Speed of estimation in positron emission tomography and related inverse problems, Annals of Statistics 18, 251-280.

Katkovnik,V.Ya. (1979) Linear and nonlinear methods for nonparametric regression analysis. Avtomatika, No.5, 35-46.

Katkovnik,V.Ya. (1983) Convergence of linear and nonlinear nonparametric kernel estimates. Avtomatika i Telemekhanika, No.4, 108-120.

Katkovnik,V.Ya. (1985) Nonparametric Identification and Data Smoothing: Local Approximation Approach. Moscow: Nauka (in Russian).

Khasminskii,R.Z. and Lebedev,V.S. (1990) On the properties of parametric estimators for areas of a discontinuous image. Probl. of Control and Information Theory, 19 (5-6), 375-385.

Kolmogorov,A.N. and Tikhomirov,V.M. (1961) ε-entropy and ε-capacity of sets in functional spaces. Amer.Math.Soc.Transl. (ser.2) 17, 277-364.

Korostelev,A.P. (1987a) On minimax estimation of a discontinuous signal, Theory of Probability and its Applications 32, 796-799.

Korostelev,A.P. (1987b) Asymptotically minimax filtering of nonparametric signal. Preprint, Inst. for System Studies, Moscow (in Russian).

Korostelev,A.P. (1988) Minimax image reconstruction problem(Abstract). In: Statistical Control Problems, Vol.83, Vilnius, Inst. Math. Cybernetics, Lithuanian Acad. Sci., 235-239 (in Russian).

Korostelev,A.P. (1991) Minimax reconstruction of two-dimensional images, Theory of Probability and its Applications 36, 153-159.

Korostelev,A.P. and Tsybakov,A.B. (1989) The minimax accuracy of image reconstruction in the problem of tomography. In: Fifth Vilnius Conf. Probab. Th. Statist. Abstracts,Vol.1,270-271.

Korostelev,A.P. and Tsybakov,A.B. (1990) Minimax nonparametric estimation of boundaries on images. In: Nonparametric and Robust Statistical Methods in Cybernetics and Informatics. Part 1. Tomsk: Tomsk State Univ., 258-262 (in Russian).

Korostelev,A.P. and Tsybakov,A.B. (1991) Optimal rates of convergence of estimators in a probabilistic setup of tomography problem, Problems of Information Transmission, 27, 73-81.

Korostelev,A.P. and Tsybakov,A.B. (1992a) Asymptotically minimax image reconstruction problems. In: Topics in Nonparametric Estimation, Khasminskii R.Z., ed., AMS, Providence, RI, 45-86.

Korostelev,A.P. and Tsybakov,A.B. (1992b) Minimax linewise algorithm for image reconstruction. CORE Discussion Paper n.9249.

Korostelev,A.P. and Tsybakov,A.B. (1992c) Estimation of support of aprobability density and estimation of support functionals. CORE Discussion Paper n.9229.

Lee,J.S. (1983) Digital image smoothing and the sigma-filter. Computer Vision, Graphics and Image Processing, 24, 255-269.

Mammen,E. and Tsybakov,A.B. (1992) Asymptotical minimax results in image analysis for sets with smooth boundaries. Institute of Statistics Discussion paper n.9205, Université Catholique de Louvain.

Marr,D. (1982) Vision. W.H.Freeman and Co., San Francisco.

Minsky,M. and Papert,S. (1969) Perceptrons. MIT Press, Cambridge, MA.

Moore,M. (1984) On the estimation of a convex set. Ann. Statist., 12, 1090-1099.

Mumford,D. and Shah,J. (1989) Optimal approximation by piecewise smooth functions and associated variational problems. Comm. Pure Appl. Math. 42, 577-685.

Müller,D.W. and Sawitzki,G. (1987) Using excess mass estimates to investigate the modality of a distribution. Preprint No. 398, SFB 123, Universität Heidelberg.

Müller,D.W. and Sawitzki,G. (1991) Excess mass estimates and tests for multimodality. J. of the American Statistical Association 86, n.415, 738-746.

Müller,H.-G. (1987) Weighted local regression and kernel methods for nonparametric curve fitting. J. of the American Statistical Association 82, n.397, 231-239.

Müller,H.-G. and Song,K.-S. (1992) On the estimation of multidimensional boundaries. Unpublised manuscript.

Nagao,M. and Matsujama,T. (1979) Edge preserving smoothing. Comput. Graphics and Image Processing 9, 394-407.

Natterer,F. (1980) A Sobolev space analysis of picture reconstruction, SIAM J. Appl. Math. 39, 402-411.

Natterer,F. (1986) The Mathematics of Computerized Tomography, Teubner, Stuttgart; J.Wiley, Chichester.

Nemirovskii,A.S. (1985) On nonparametric estimation of smooth regression functions, Izv. Acad. Nauk SSSR, Techn. Kibern. No.3, 50-60.

Nemirovskii,A.S. (1990) On necessary conditions for the efficient estimation of functionals of a nonparametric signal which is observed in white noise, Theory of Probability and its Applications 35, 83-91.

Polonik,W. (1991) Excess mass estimates over classes of sets, estimation of the density contour clusters and tests for multimodality in more than one dimensions. Preprint No. 611, SFB 123, Universität Heidelberg.

Pratt,W.K. (1978) Digital Image Processing, J.Wiley, N.Y.

Rényi,A. and Sulanke,R. (1963) Über die konvexe Hülle von n zufällig gewählten Punkten. Z. Wahrscheinlichkeitstheorie verw. Geb. 2, 75-84.

Rényi,A. and Sulanke,R. (1964) Über die konvexe Hülle von n zufällig gewählten Punkten.II. Z. Wahrscheinlichkeitstheorie verw. Geb. 3, 138-147.

Ripley,B.D. (1988) Statistical Inference for Spatial Processes. Cambridge Univ. Press.

Ripley,B.D. and Rasson,J.P. (1977) Finding the edge of a Poisson forest. J. Appl. Prob., 14, 483-491.

Rosenfeld,A. and Kak,A.C. (1982) Digital Picture Processing. Academic Press, London.

Rudemo,M., Skovgaard,I. and Stryhn,H. (1990) Maximum likelihood estimation of curves in images. Report 90-4, Dept.of Math. and Phys., Royal Veterinary and Agricultural Univ., Denmark.

Rudemo,M. and Stryhn,H. (1991) Approximating the distribution of maximum likelihood contour estimators in two-region images. Report 91-2, Dept.of Math. and Phys., Royal Veterinary and Agricultural Univ., Denmark.

Schneider,R. (1988) Random approximation of convex sets. J. of Microscopy, 151, part 3, 211-227.

Shiau, J.J., Wahba,G. and Johnson,D.R. (1986) Partial spline models for the inclusion of tropopause and frontal boundary information in otherwise smooth two- and three-dimensional objective analysis. J. Atmospheric and Oceanic Technology 3, 714-725.

Shiryaev,A.N. (1978) Optimal Stopping Rules. Springer, N.Y. e.a.

Siegmund,D. (1985) Sequential Analysis: Test and Confidence Intervals. Springer, N.Y. e.a.

Stone,C.J. (1977) Consistent nonparametric regression (with discussion). Annals of Statistics 5, 595-645.

Stone,C.J. (1980) Optimal rates of convergence for nonparametric estimators, Annals of Statistics 8, 1348-1360.

Stone,C.J. (1982) Optimal global rates of convergence for nonparametric estimators, Annals of Statistics 10, 1040-1053.

Titterington,D.M. (1985) Common structure of smoothing techniques in statistics. Internat. Statist. Review 53, 141-170.

Torre,V. and Poggio,T. (1986) On edge detection. IEEE Trans. on PAMI, 8, 147-163.

Tsybakov,A.B. (1989) Optimal estimation accuracy of nonsmooth images,Problems of Information Transmission,25,180-191.

Tsybakov,A.B. (1991) Nonparametric techniques in image estimation. In: Nonparametric Functional Estimation and Related Topics, Proceedings NATO ASI, Spetses (Greece), G.Roussas, ed., 669-678.

Tsybakov,A.B. (1992) Multidimensional change-point problems and boundary estimation. Institute of Statistics Discussion paper n.92011, Université Catholique de Louvain.

Yosida,K. (1965) Functional Analysis. Springer, Berlin e.a.

AUTHOR INDEX

SUBJECT INDEX

Vol. 1: R.A. Fisher: An Appreciation. Edited by S.E. Fienberg and D.V. Hinkley. XI, 208 pages, 1980.

Vol. 2: Mathematical Statistics and Probability Theory. Proceedings 1978. Edited by W. Klonecki, A. Kozek, and J. Rosinski. XXIV, 373 pages, 1980.

Vol. 3: B.D. Spencer, Benefit-Cost Analysis of Data Used to Allocate Funds. VIII, 296 pages, 1980.

Vol. 4: E.A. van Doorn, Stochastic Monotonicity and Queueing Applications of Birth-Death Processes. VI, 118 pages, 1981.

Vol. 5: T. Rolski, Stationary Random Processes Associated with Point Processes. VI, 139 pages, 1981.

Vol. 6: S.S. Gupta and D.-Y. Huang, Multiple Statistical Decision Theory: Recent Developments. VIII, 104 pages, 1981.

Vol. 7: M. Akahira and K. Takeuchi, Asymptotic Efficiency of Statistical Estimators. VIII, 242 pages, 1981.

Vol. 8: The First Pannonian Symposium on Mathematical Statistics. Edited by P. Révész, L. Schmetterer, and V.M. Zolotarev. VI, 308 pages, 1981.

Vol. 9: B. Jørgensen, Statistical Properties of the Generalized Inverse Gaussian Distribution. VI, 188 pages, 1981.

Vol. 10: A.A. McIntosh, Fitting Linear Models: An Application of Conjugate Gradient Algorithms. VI, 200 pages, 1982.

Vol. 11: D.F. Nicholls and B.G. Quinn, Random Coefficient Autoregressive Models: An Introduction. V, 154 pages, 1982.

Vol. 12: M. Jacobsen, Statistical Analysis of Counting Processes. VII, 226 pages, 1982.

Vol. 13: J. Pfanzagl (with the assistance of W. Wefelmeyer), Contributions to a General Asymptotic Statistical Theory. VII, 315 pages, 1982.

Vol. 14: GLIM 82: Proceedings of the International Conference on Generalised Linear Models. Edited by R. Gilchrist. V, 188 pages, 1982.

Vol. 15: K.R.W. Brewer and M. Hanif, Sampling with Unequal Probabilities. IX, 164 pages, 1983.

Vol. 16: Specifying Statistical Models: From Parametric to Non-Parametric, Using Bayesian or Non-Bayesian Approaches. Edited by J.P. Florens, M. Mouchart, J.P. Raoult, L. Simar, and A.F.M. Smith, XI, 204 pages, 1983.

Vol. 17: I.V. Basawa and D.J. Scott, Asymptotic Optimal Inference for Non-Ergodic Models. IX, 170 pages, 1983.

Vol. 18: W. Britton, Conjugate Duality and the Exponential Fourier Spectrum. V, 226 pages, 1983.

Vol. 19: L. Fernholz, von Mises Calculus For Statistical Functionals. VIII, 124 pages, 1983.

Vol. 20: Mathematical Learning Models — Theory and Algorithms: Proceedings of a Conference. Edited by U. Herkenrath, D. Kalin, W. Vogel. XIV, 226 pages, 1983.

Vol. 21: H. Tong, Threshold Models in Non-linear Time Series Analysis. X, 323 pages, 1983.

Vol. 22: S. Johansen, Functional Relations, Random Coefficients and Nonlinear Regression with Application to Kinetic Data, VIII, 126 pages, 1984.

Vol. 23: D.G. Saphire, Estimation of Victimization Prevalence Using Data from the National Crime Survey. V, 165 pages, 1984.

Vol. 24: T.S. Rao, M.M. Gabr, An Introduction to Bispectral Analysis and Bilinear Time Series Models. VIII, 280 pages, 1984.

Vol. 25: Time Series Analysis of Irregularly Observed Data. Proceedings, 1983. Edited by E. Parzen. VII, 363 pages, 1984.

Vol. 26: Robust and Nonlinear Time Series Analysis. Proceedings, 1983. Edited by J. Franke, W. Härdle and D. Martin. IX, 286 pages, 1984.

Vol. 27: A. Janssen, H. Milbrodt, H. Strasser, Infinitely Divisible Statistical Experiments. VI, 163 pages, 1985.

Vol. 28: S. Amari, Differential-Geometrical Methods in Statistics. V, 290 pages, 1985.

Vol. 29: Statistics in Ornithology. Edited by B.J.T. Morgan and P.M. North. XXV, 418 pages, 1985.

Vol 30: J. Grandell, Stochastic Models of Air Pollutant Concentration. V, 110 pages, 1985.

Vol. 31: J. Pfanzagl, Asymptotic Expansions for General Statistical Models. VII, 505 pages, 1985.

Vol. 32: Generalized Linear Models. Proceedings, 1985. Edited by R. Gilchrist, B. Francis and J. Whittaker. VI, 178 pages, 1985.

Vol. 33: M. Csörgo, S. Csörgo, L. Horváth, An Asymptotic Theory for Empirical Reliability and Concentration Processes. V, 171 pages, 1986.

Vol. 34: D.E. Critchlow, Metric Methods for Analyzing Partially Ranked Data. X, 216 pages, 1985.

Vol. 35: Linear Statistical Inference. Proceedings, 1984. Edited by T. Calinski and W. Klonecki. VI, 318 pages, 1985.

Vol. 36: B. Matérn, Spatial Variation. Second Edition. 151 pages, 1986.

Vol. 37: Advances in Order Restricted Statistical Inference. Proceedings, 1985. Edited by R. Dykstra, T. Robertson and F.T. Wright. VIII, 295 pages, 1986.

Vol. 38: Survey Research Designs: Towards a Better Understanding of Their Costs and Benefits. Edited by R.W. Pearson and R.F. Boruch. V, 129 pages, 1986.

Vol. 39: J.D. Malley, Optimal Unbiased Estimation of Variance Components. IX, 146 pages, 1986.

Vol. 40: H.R. Lerche, Boundary Crossing of Brownian Motion. V, 142 pages, 1986.

Vol. 41: F. Baccelli, P. Brémaud, Palm Probabilities and Stationary Queues. VII, 106 pages, 1987.

Vol. 42: S. Kullback, J.C. Keegel, J.H. Kullback, Topics in Statistical Information Theory. IX, 158 pages, 1987.

Vol. 43: B.C. Arnold, Majorization and the Lorenz Order: A Brief Introduction. VI, 122 pages, 1987.

Vol. 44: D.L. McLeish, Christopher G. Small, The Theory and Applications of Statistical Inference Functions. VI, 124 pages, 1987.

Vol. 45: J.K. Ghosh (Ed.), Statistical Information and Likelihood. 384 pages, 1988.

Vol. 46: H.-G. Müller, Nonparametric Regression Analysis of Longitudinal Data. VI, 199 pages, 1988.

Vol. 47: A.J. Getson, F.C. Hsuan, {2}-Inverses and Their Statistical Application. VIII, 110 pages, 1988.

Vol. 48: G.L. Bretthorst, Bayesian Spectrum Analysis and Parameter Estimation. XII, 209 pages, 1988.

Vol. 49: S.L. Lauritzen, Extremal Families and Systems of Sufficient Statistics. XV, 268 pages, 1988.

Vol. 50: O.E. Barndorff-Nielsen, Parametric Statistical Models and Likelihood. VII, 276 pages, 1988.

Vol. 51: J. Hüsler, R.-D. Reiss (Eds.), Extreme Value Theory. Proceedings, 1987. X, 279 pages, 1989.

Vol. 52: P.K. Goel, T. Ramalingam, The Matching Methodology: Some Statistical Properties. VIII, 152 pages, 1989.

Vol. 53: B.C. Arnold, N. Balakrishnan, Relations, Bounds and Approximations for Order Statistics. IX, 173 pages, 1989.

Vol. 54: K.R. Shah, B.K. Sinha, Theory of Optimal Designs. VIII, 171 pages, 1989.

Vol. 55: L. McDonald, B. Manly, J. Lockwood, J. Logan (Eds.), Estimation and Analysis of Insect Populations. Proceedings, 1988. XIV, 492 pages, 1989.

Vol. 56: J.K. Lindsey, The Analysis of Categorical Data Using GLIM. V, 168 pages, 1989.

Vol. 57: A. Decarli, B.J. Francis, R. Gilchrist, G.U.H. Seeber (Eds.), Statistical Modelling. Proceedings, 1989. IX, 343 pages, 1989.

Vol. 58: O.E. Barndorff-Nielsen, P. Blæsild, P.S. Eriksen, Decomposition and Invariance of Measures, and Statistical Transformation Models. V, 147 pages, 1989.

Vol. 59: S. Gupta, R. Mukerjee, A Calculus for Factorial Arrangements. VI, 126 pages, 1989.

Vol. 60: L. Györfi, W. Härdle, P. Sarda, Ph. Vieu, Nonparametric Curve Estimation from Time Series. VIII, 153 pages, 1989.

Vol. 61: J. Breckling, The Analysis of Directional Time Series: Applications to Wind Speed and Direction. VIII, 238 pages, 1989.

Vol. 62: J.C. Akkerboom, Testing Problems with Linear or Angular Inequality Constraints. XII, 291 pages, 1990.

Vol. 63: J. Pfanzagl, Estimation in Semiparametric Models: Some Recent Developments. III, 112 pages, 1990.

Vol. 64: S. Gabler, Minimax Solutions in Sampling from Finite Populations. V, 132 pages, 1990.

Vol. 65: A. Janssen, D.M. Mason, Non-Standard Rank Tests. VI, 252 pages, 1990.

Vol. 66: T. Wright, Exact Confidence Bounds when Sampling from Small Finite Universes. XVI, 431 pages, 1991.

Vol. 67: M.A. Tanner, Tools for Statistical Inference: Observed Data and Data Augmentation Methods. VI, 110 pages, 1991.

Vol. 68: M. Taniguchi, Higher Order Asymptotic Theory for Time Series Analysis. VIII, 160 pages, 1991.

Vol. 69: N.J.D. Nagelkerke, Maximum Likelihood Estimation of Functional Relationships. V, 110 pages, 1992.

Vol. 70: K. Iida, Studies on the Optimal Search Plan. VIII, 130 pages, 1992.

Vol. 71: E.M.R.A. Engel, A Road to Randomness in Physical Systems. IX, 155 pages, 1992.

Vol. 72: J.K. Lindsey, The Analysis of Stochastic Processes using GLIM. VI, 294 pages, 1992.

Vol. 73: B.C. Arnold, E. Castillo, J.-M. Sarabia, Conditionally Specified Distributions. XIII, 151 pages, 1992.

Vol. 74: P. Barone, A. Frigessi, M. Piccioni, Stochastic Models, Statistical Methods, and Algorithms in Image Analysis. VI, 258 pages, 1992.

Vol. 75: P.K. Goel, N.S. Iyengar (Eds.), Bayesian Analysis in Statistics and Econometrics. XI, 410 pages, 1992.

Vol. 76: L. Bondesson, Generalized Gamma Convolutions and Related Classes of Distributions and Densities. VIII, 173 pages, 1992.

Vol. 77: E. Mammen, When Does Bootstrap Work? Asymptotic Results and Simulations. VI, 196 pages, 1992.

Vol. 78: L. Fahrmeir, B. Francis, R. Gilchrist, G. Tutz (Eds.), Advances in GLIM and Statistical Modelling: Proceedings of the GLIM92 Conference and the 7th International Workshop on Statistical Modelling, Munich, 13-17 July 1992. IX, 225 pages, 1992.

Vol. 79: N. Schmitz, Optimal Sequentially Planned Decision Procedures. XII, 209 pages, 1992.

Vol. 80: M. Fligner, J. Verducci (Eds.), Probability Models and Statistical Analyses for Ranking Data. XXII, 306 pages, 1992.

Vol. 81: P. Spirtes, C. Glymour, R. Scheines, Causation, Prediction, and Search. XXIII, 526 pages, 1993.

Vol. 82: A. P. Korostelev and A. B. Tsybakov, Minimax Theory of Image Reconstruction. XII, 268 pages, 1993.

General Remarks

Lecture Notes are printed by photo-offset from the master-copy delivered in camera-ready form by the authors of monographs, resp. editors of proceedings volumes. For this purpose Springer-Verlag provides technical instructions for the preparation of manuscripts. Volume editors are requested to distribute these to all contributing authors of proceedings volumes. Some homogeneity in the presentation of the contributions in a multi-author volume is desirable.

Careful preparation of manuscripts will help keep production time short and ensure a satisfactory appearance of the finished book. The actual production of a Lecture Notes volume normally takes approximately 8 weeks.

For monograph manuscripts typed or typeset according to our instructions, Springer-Verlag can, if necessary, contribute towards the preparation costs at a fixed rate.

Authors of monographs receive 50 free copies of their book. Editors of proceedings volumes similarly receive 50 copies of the book and are responsible for redistributing these to authors etc. at their discretion. No reprints of individual contributions can be supplied. No royalty is paid on Lecture Notes volumes.

Volume authors and editors are entitled to purchase further copies of their book for their personal use at a discount of 33.3% and other Springer mathematics books at a discount of 20% directly from Springer-Verlag. Authors contributing to proceedings volumes may purchase the volume in which their article appears at a discount of 20%.

Springer-Verlag secures the copyright for each volume.

GPSR Compliance

The European Union's (EU) General Product Safety Regulation (GPSR) is a set of rules that requires consumer products to be safe and our obligations to ensure this.

If you have any concerns about our products, you can contact us on ProductSafety@springernature.com

In case Publisher is established outside the EU, the EU authorized representative is:

Springer Nature Customer Service Center GmbH
Europaplatz 3
69115 Heidelberg, Germany

Batch number: 10704713

Printed by Printforce, the Netherlands